W9-BZV-698

Fortifications

Afghanistan Cave Complexes 1979–2004

Mountain strongholds of the Mujahideen, Taliban & Al Qaeda

Mir Bahmanyar • Illustrated by Ian Palmer
Series editors Marcus Cowper and Nikolai Bogdanovic

First published in Great Britain in 2004 by Osprey Publishing, Elms Court, Chapel Way, Botley, Oxford OX2 9LP, United Kingdom.
Email: info@ospreypublishing.com

ISBN 1 84176 958 4

SERIES EDITORS: Marcus Cowper and Nikolai Bogdanovic

Editor: Ilios Publishing, Oxford, UK (www.iliospublishing.com)
Design: Ken Vail Graphic Design, Cambridge, UK
Index by David Worthington
Maps by The Map Studio Ltd
Originated by Grasmere Digital Imaging, Leeds, UK
Printed and bound by L-Rex Printing Company Ltd

04 05 06 07 08 10 9 8 7 6 5 4 3 2 1

A CIP catalog record for this book is available from the British Library.

FOR A CATALOG OF ALL BOOKS PUBLISHED BY OSPREY MILITARY AND AVIATION PLEASE CONTACT:

Osprey Direct UK, PO Box 140, Wellingborough, Northants, NN8 2FA, United Kingdom.
Email: info@ospreydirect.co.uk

Osprey Direct USA, c/o MBI Publishing, PO Box 1, 729 Prospect Ave, Osceola, WI 54020, USA.
Email: info@ospreydirectusa.com

www.ospreypublishing.com

Acknowledgments

Special thanks to Sara van Valkenburg and Lester Grau of FMSO; Chris Osman of Tactical Assault Gear; Karin Martinez, 10th Mountain Division PAO; Major Hugh Cate III, 101st Airborne Division PAO; Major Matt Morgan, USMC PAO; and Val Webb and Mark Brender from Space Imaging.

The Fortress Study Group (FSG)

The object of the FSG is to advance the education of the public in the study of all aspects of fortifications and their armaments, especially works constructed to mount or resist artillery. The FSG holds an annual conference in September over a long weekend with visits and evening lectures, an annual tour abroad lasting about eight days, and an annual Members' Day.

The FSG journal *FORT* is published annually, and its newsletter *Casemate* is published three times a year. Membership is international. For further details, please contact:

The Secretary, c/o 6 Lanark Place, London W9 1BS, UK

Contents

Introduction

The peoples of Afghanistan have been able to utilize their massive mountain ranges and caves to repel numerous onslaughts from foreign invaders for centuries. In ancient times, militant Assyria was unable to subdue the independent tribes of the area. The hugely successful armies of Alexander the Great were bogged down in mountain warfare for three years. The highly mobile and ruthless Mongol invasions of the 13th century met with no more success then the highly mechanized armies of the Soviets seven centuries later. The United States invaded Afghanistan in 2001 in order to oust the Taliban regime and crush the Al Qaeda network held responsible for numerous attacks throughout the world. Though successful in the initial invasion, the most sophisticated and best-supplied military the world has ever witnessed has been unable to control the country and its different inhabitants. Various groups, including fighters from the Al Qaeda and Taliban organizations, have used the age-old mountains and their caves to withstand the tremendous weaponry of the Coalition forces for the past three years. Irrigation channels, mountain strongholds, caves and tunnel systems dot the countryside, and have proven to be unassailable in the long term. Not all of the natural or manmade defenses are identical. For example, Tora Bora (Black Dust), one of the first areas bombed by Coalition forces where Al Qaeda's most prominent member Osama Bin Laden was thought to have hidden, is a series of individual caves, whereas the Zhawar region is much more developed. Some sites are logistics depots or supply points and vary from complex mazes of tunnels to simple cliff overhangs. Taken together, however, the cave and tunnel systems of Afghanistan are difficult obstacles for any invading army to overcome.

Technological advances over the centuries have changed the nature of warfare, but age-old maxims still hold true in the 21st century, especially in the unforgiving mountains of Afghanistan. Soldiers and sophisticated equipment cannot operate effectively at high altitudes for long periods of time, and the simplest of tasks can become extremely difficult. Exhaustion and oxygen deprivation are just some of the obstacles the modern infantry grunt must overcome to fight effectively. Helicopters and fast-moving fixed-wing aircraft are only as effective as the terrain and weather conditions permit. High-altitude carpet bombing, smart bombs and thermobaric (heat and pressure) weapons, combined with the ability to operate effectively at night, have clearly given an edge to the Coalition soldier. Thousands of guerrillas and civilians have been killed, and there have been, as in all other wars, friendly fire casualties, but no matter; the incredibly rugged and beautiful yet desolate mountain ranges dominate the war, and render mankind seemingly insignificant. Mountainous regions in warfare have always sided with the defenders and never the invading army.

Another interesting aspect to the conflict in Afghanistan is the absence of a unified military command

The mountainous region of Kohe Sofi on March 27, 2003. Afghanistan has been repeatedly searched by elements of the 82nd Airborne Division for weapons and fighters. The terrain is a natural obstacle to conventional war fighting. (U.S. Army)

structure. Indeed the various ethnolinguistic tribes, warlords and political parties have fought each other with the same dedication that they demonstrated toward destroying the will of the Soviets from 1979 to 1989. This disparity, this lack of a central military authority, coupled with the extreme geography, makes the conquest of Afghanistan impossible; for how can any army pin down and destroy conventional units that don't exist? Militias disperse and unite according to seemingly intangible rules, and at times, contrary to military doctrine, amid the mountains and caves that serve as ideal places to weather the foreign and domestic military storms.

It is very little surprise then that the war in Afghanistan cannot be won by killing large numbers of the enemy, although this may be helpful and certainly allows for body count statisticians to announce victories. But "winning" the war in Afghanistan is, like all wars before, dependent on destroying the will of the people. And as the mountains bear witness, the will of the Afghani peoples is forged by centuries of deprivation and hardship. According to the International Red Cross, mines and unexploded ordnance in Afghanistan have injured more than 200,000 people over the last two decades of war. The country is one of the most heavily mined nations on Earth. Nearly 1,000,000 Afghanis were killed during the 20 years of conflict in the latter half of the 20th century, including 3,000 killed by toxic agents, and millions more are living as refugees throughout the neighboring nation-states. Islam is the only common thread that provides some stability within the war-torn country.

Marines from E Company, 2nd Battalion, 8th Marines, pull security on the Naghlu Bridge in Sarobi on April 13, 2004. (U.S. Army)

The country

Afghanistan is a landlocked country consisting of three basic geographical regions and covers an area of 250,985 square miles (650,000km^2), approximately the size of the state of Texas. To the north, Afghanistan's neighbors are Turkmenistan, Uzbekistan, and Tajikistan. Pakistan lies to the east and south, and Iran to the west. In the extreme northeast lies a long, narrow panhandle, the Wakhan Corridor.

Three distinct geographical regions mark the country. The Northern Plains (Steppe) are characterized by mountainous plateaus and rolling hills at an average elevation of 2,000ft above sea level. The plains are actually steppes with seasonal grasslands capable of supporting a small nomadic population. The Central (Afghan) Highlands cover two-thirds of the country and has elevations up to 25,000ft. Finally the Southwestern Plateau (Desert Basins) is a mostly arid environment; a barren desert with large areas of drifting sand, scattered hill belts, and a few low mountains. Its elevation ranges from 20,000ft in the east, to 500ft above sea level in the west.

The Hindu Kush (Hindu Killer) is a spectacular mountain range, 1,000 miles long and 200 miles wide, running northeast to southwest. Subsidiary ranges continue south and west with decreasing elevations, gradually merging into the plains that continue into Iran and Pakistan. It has over two-dozen summits that measure 24,000ft (7,315m) or higher. In the east, the mountains are indistinguishable from those of Pakistan. The snow-capped Hindu Kush have sharp-crested ridges and towering peaks, while the lower, western mountains are generally rounded or flat-topped. Historically, the passes across the Hindu Kush, in particular the Khyber Pass, have been of great military significance, providing access to the northern plains of India.

ABOVE LEFT A U.S. Navy SEAL clears a crevice/cave entrance in eastern Afghanistan. Numerous natural overhangs or crevices can be used as is or further improved upon. No matter which, these areas are difficult obstacles to overcome for any foreign army. (Tactical Assault Gear)

ABOVE RIGHT 2nd Battalion, 27th Infantry, 25th Infantry Division (Light), soldiers pass through a valley during a mission on April 5, 2004. (U.S. Army)

Afghanistan is mostly dry with seasonal temperature extremes. Countrywide, the extreme summer high temperature is 118°F in the west and the extreme winter low is –4°F in Kabul. The rainy season, usually scant and drought-plagued, lasts from October through April, and extends across the entire winter, which lasts from December through February. Summer runs from June through August and is usually sunny, dry and hot.

Afghanistan has four major river systems that originate in the Hindu Kush: the Kabul, the Helmand, the Amu Darya and the Harirud. Of the four, only the eastward-flowing Kabul reaches the ocean; the other three eventually disappear into salt marshes or desert wastes.

All the Afghan rivers and their tributaries are used for irrigation. Supplementing the stream irrigation is the *karez*, a system of underground channels (with vertical access and maintenance shafts) carrying water from the base of the mountain slopes to oases on the valley floors. The signature of *karez* (*qanat* in Iran), particularly noticeable from the air, is the row of evenly spaced openings (shafts) surrounded by mounds of earth that define the course of the underground channels.

The U.S. Library of Congress study on Afghanistan states that Pashtuns are the country's dominant ethnic group, comprising about 38 percent of the population. Tajik (25 percent), Hazara (19 percent), Uzbek (6 percent), Aimaq, Turkmen, Baluch and other small groups also are represented. Dari (Afghan-Persian) and Pashto are official languages. Dari is spoken by more than one-third of the population as a first language and serves as a lingua franca for

RIGHT The view from the crew door of a UH-60Q Blackhawk, from the 717th Medical Company (Air Ambulance), shows the rough terrain of northeastern Afghanistan on the route back from Salerno Forward Operating Base (FOB) on January 25, 2004. (U.S. Army)

LEFT A UH-60Q Blackhawk from the 717th Medical Company (Air Ambulance) flies through the scenic mountains of the Gardez Pass returning from Salerno Forward Operating Base (FOB) after an equipment change over on January 25, 2004. (U.S. Army)

most Afghans, though the Taliban use Pashto. Tajik, Uzbek, and Turkmen are spoken widely in the north. Various smaller groups throughout the country speak more than 70 other languages and numerous dialects. Afghanistan is an Islamic country, with an estimated 84 percent of the population Sunni, and the remainder predominantly Shi'a. Throughout the civil war years that ravaged Afghanistan, the division tended to fall along ethnic lines as the Pashtuns sided with Gulabudin Hikmatyar's Islamic Party (Hizb-e Islami), a Pashtun party, while the non-Pashtuns supported Ahmed Shah Massoud, a Tajik and Burhanuddin Rabani's military commander, and his Supervisory Council (Shoray-e Nezar)/Islamic Society (Jamiyat-e Islami). There are hundreds of Islamic groups throughout the region, all vying for power. Ethnic cleansing, which has most recently been associated with Eastern Europe and parts of

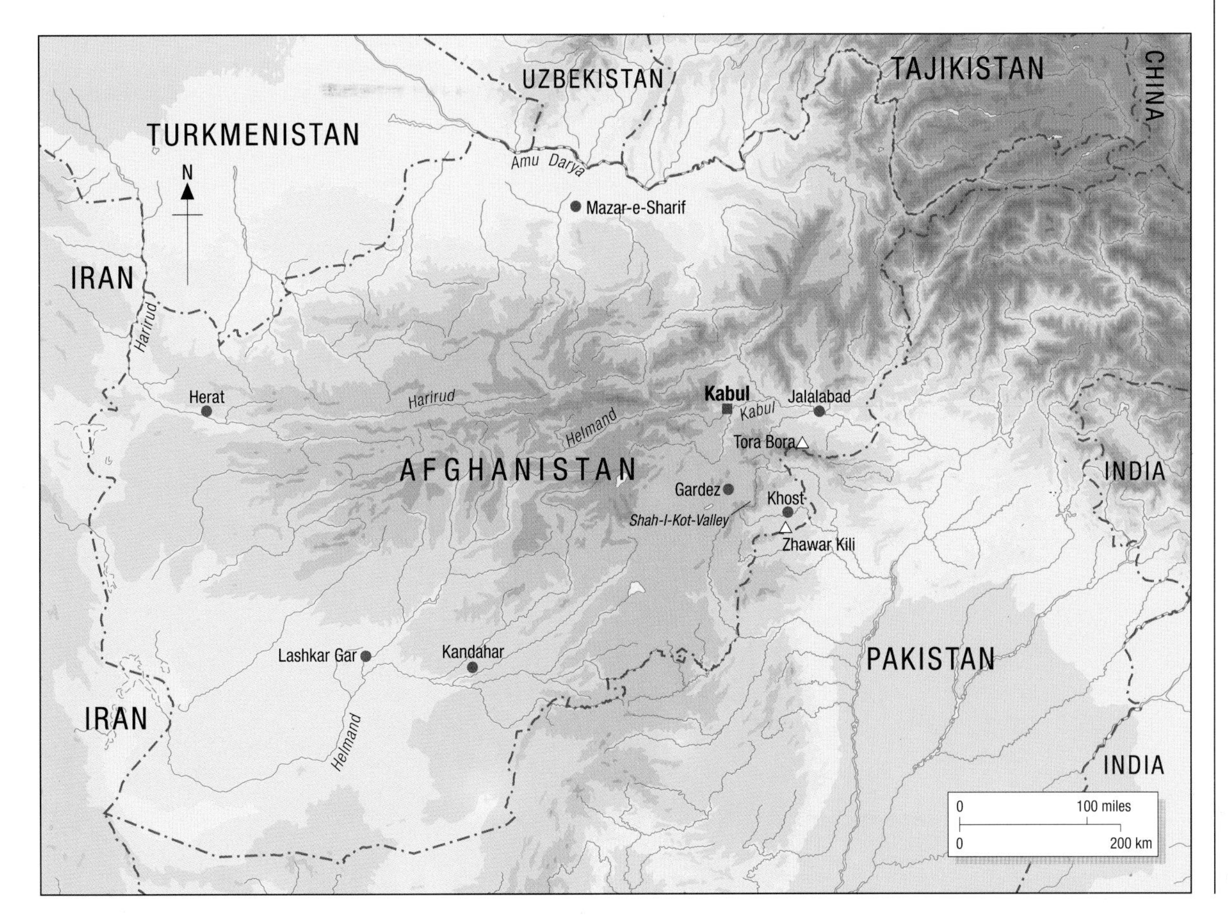

Afghanistan is a land-locked country with massive mountain ranges in the east, which have been used extensively over the centuries as bulwarks against invading forces. (© Copyright Osprey Publishing Ltd)

RIGHT An aerial view of a destroyed Russian tank located on a hilltop near Bagram Airfield, December 15, 2003. Huge amounts of destroyed Russian military equipment remain scattered throughout Afghanistan, and serve as a reminder of the fierce battles that raged here during the Soviet occupation. (U.S. Army)

BELOW On Saturday November 15, 2003, a Romanian priest and the bearer of a cross lead a procession for a fellow soldier on Bagram Airfield. This procession is held to honor fallen Romanian soldier Anton Mihail Samuila, who lost his life during Operation Enduring Freedom. (U.S. Army)

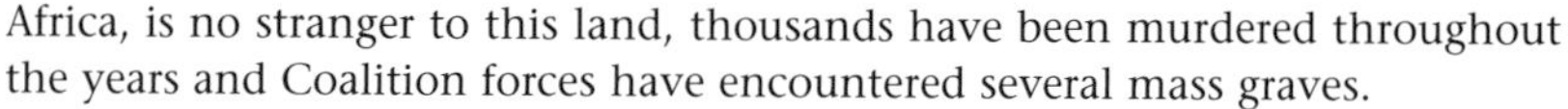

Africa, is no stranger to this land, thousands have been murdered throughout the years and Coalition forces have encountered several mass graves.

Afghanistan is one of the world's poorest and least developed countries. The infant mortality rate based on a 2000 estimate is 149:1,000 births, and life expectancy is approximately 47 for men and 45 for women. Key cities are Kandahar, Kabul, Jalalabad, Mazar-e-Sharif and Herat.

Although Afghanistan yields natural resources, such as natural gas, oil, coal, copper, chromite, talc, barites, sulfur, lead, zinc, iron, salt, precious and semiprecious stones, it is still an agrarian society producing wheat, corn, barley, rice, cotton, fruit, nuts, wool and mutton. In addition, exports include gems and carpets. Afghanistan produces 80 percent of the world's opium supply with estimates ranging up to $2.5 billion in underground revenue in 2002–03. Farmers have no choice but to raise poppy plants.

Chronology

150 BC–AD 700 Central Asian and Sassanian rule. Buddhism introduced

AD 700–900 Islamic Conquest

1220–1506 Mongol conquest and rule

1500–1747 Mughal-Safavid rivalry. Persian Nadir Shah replaces Mughal dynasty

1747–1818 Afghan tribal leader Ahmad Shah creates the Durrani dynasty, which continues through most of modern-day Afghanistan.

1825–79 The "Great Game" – the rise of Dost Mohammad during the Anglo-Russian struggle for control of Afghanistan

1838–42 The First Anglo-Afghan War caused by the British attempt to replace Dost Mohammad

1878 The Second Anglo-Afghan War breaks out. Causes include Russian encroachment and British pressures

1880–1901 Abdr Rahman Khan, "The Iron Amir." Borders redrawn vis-à-vis Russia, British India and Persia

1901–19 The Reign Of King Habibullah

1907 Afghanistan becomes independent, British influence remains over foreign affairs

1919 Third Anglo-Afghan War and Independence. Afghanistan invades India to regain control over foreign affairs

1919–29 The reign of King Amanullah

1929–33 Rule Of Muhammad Nadir Shah. His cautious modernization efforts lead to his assassination

1933–73 Rule of Mohammad Zahir Shah

1937 Treaty of Saadabad reinforces ties with neighboring Islamic states, Iran, Iraq, and Turkey

1946 Afghanistan joins United Nations

1953–63 Mohammad Daoud, cousin of Mohammad Zahir, becomes Prime Minister. His term is complicated with issues of internal strife and Cold War politics, Pakistani agitations, and the partitioning of British India; Pakistan routinely interferes in Afghanistan politics. Daoud resigns in 1963

1965 The People's Democratic Party of Afghanistan (PDPA) is founded. It is a communist party comprising a small group of men, followers of Nur Mohammad Taraki and Babrak Karmal, both avowed Marxist-Leninists with a pro-Moscow slant

1967 PDPA splits into several factions. The two most important of these are the Khalq (Masses) faction, headed by Taraki, and the Parcham (Banner) faction, headed by Karmal

1973–78 Mohammad Daoud stages coup d'etat and proclaims republic

January 4, 1978 First communist president takes control of Afghanistan in a bloody coup that results in the death of Daoud and most of the Royal family

1979 80,000 Soviet troops pour over the Afghan border to replace the communist leader with the pro-Moscow Babrak Karmal. Invasion eventually sends 5 million Afghan refugees into neighboring Iran and Pakistan

ABOVE On June 2, 2003, soldiers assigned to the 2nd Battalion, 505th Parachute Infantry Regiment and the 3rd Infantry Division, Military Police Company, secure, separate and silence Afghan individuals so that they can be questioned about terrorist activities in Shah-I-Kot, Afghanistan. (U.S. Army)

LEFT The mountain ranges of Adi Ghar present a beautiful scenic view during Operation Mongoose on January 30, 2003. (U.S. Army)

LEFT Staff Sgt Matthew Duesbery of A Company, 2nd Battalion, 504th Parachute Infantry Regiment (PIR), 82nd Airborne Division, from Kandahar Army Airfield, points to the direction of their objective during a combat patrol in the mountains of Adi Ghar as part of Operation Mongoose on January 30, 2003. (U.S. Army)

BELOW The explosion of a cave during a combat patrol in the Adi Ghar Mountains set off by the soldiers of A Company, 307th Engineers, attached to the 82nd Airborne Division as part of Operation Mongoose on January 30, 2003. (U.S. Army)

1980s A loose alliance of Islamic rebel groups, the Mujahideen, battle Soviet troops with the help of U.S. weapons and training

January 2, 1989 Last Soviet troops withdraw from Afghanistan

1992–94 Bitter factional fighting kills at least 50,000 in Kabul, mostly civilians. Various warring groups sign four peace agreements, but fighting eventually resumes

September 1, 1996 The Taliban drive President Burhanuddin Rabbani out of Kabul, capture the capital and execute Najibullah

August 20, 1998 Khost and Jalalabad are hit by U.S. cruise missiles aimed against the stronghold of Osama Bin Laden, a Saudi exile living in Afghanistan

1999 Afghanistan produces 73 percent of the world's opium, according to the U.N. Office for Drug Control and Crime Prevention

February 13, 1999 Osama Bin Laden is reported missing by his Taliban hosts in Afghanistan

July 6, 1999 U.S. President Bill Clinton imposes financial and commercial sanctions on Afghanistan's ruling Taliban movement because of its support for terrorism suspect Osama Bin Laden

July 28, 1999 Thousands of Taliban fighters launch an offensive to crush Ahmed Shah Massoud, the last hurdle between the Islamic militia and control of the whole of Afghanistan

November 14, 1999 U.N. sanctions against Afghanistan go into force, imposed for not handing over Osama Bin Laden

May 1, 2000 The annual U.S. report on international terrorism is released and names Afghanistan as posing a major terrorist threat, in part due to its hosting of Osama Bin Laden

February 2001 Taliban leader Mohammad Omar decrees all statues in the country are being worshipped as false gods, represent an insult to Islam and should be destroyed. The order leads to the destruction of

priceless historic artifacts across the country including the world's tallest statue of an upright Buddha in Bamiyan

September 9, 2001 Afghan opposition leader Ahmed Shah Massoud is assassinated. A suicide bomber, posing as a journalist, blows himself up after gaining access to Massoud's office

September 11, 2001 Hijackers use airplanes as missiles, kill more than 3,000 people and destroy the two towers of the World Trade Center and part of the Pentagon

September 14, 2001 U.S. Congress authorizes President George W. Bush to use "all necessary and appropriate force" against the terrorists who orchestrated the September 11 attacks. The vote in the U.S. Senate is unanimous and there is only one "no" vote in the U.S. House of Representatives

September 18, 2001 The United Nations Security Council demands that the Islamic Emirate of Afghanistan "immediately and unconditionally" hand over Osama Bin Laden

October 7, 2001 The United States begins waging war in Afghanistan to flush out suspected Osama Bin Laden and Al Qaeda fighters, and to help topple the Taliban regime

December 10, 2001 After surrounding a giant cave complex in the eastern Afghan region of Tora Bora, United States and Afghan troops intercept a radio transmission that was believed to have come from Osama Bin Laden. U.S. warplanes blanket the area with bombs

January 9, 2002 U.S. airstrikes continue against a complex of caves, tunnels and buildings used as an Al Qaeda training camp at Zhawar Kili in the mountains of eastern Afghanistan

January 29, 2002 In the Adi Ghar mountain area about 14 miles north of Spin Boldak, U.S.-led Coalition forces, consisting of 300 men, identify 27 caves and clear 12 of them. The caves contain supplies such as food, water, blankets, fuel, mules, and signs that wounded men had been treated. U.S. and allied warplanes then pound the cave complex with 2,000 and 500lb bombs. In fire exchanges, at least 18 rebel fighters are killed. A U.S. AH-64 helicopter comes under small-arms fire. This is part of Operation Mongoose

February 13, 2002 In Operation Eagle Fury, Coalition warplanes drop four 500lb bombs and fire several hundred rounds of ammunition at the caves. Special forces patrols had collected abandoned ammunition casings and rocket-launchers. Fifteen fighters are captured by more than 100 U.S. troops, while an estimated 30 rebels are believed to have suffered heavy injuries

February 13, 2003 Operation Viper begins as United States CH-47 Chinook helicopters carrying U.S. troops touch down in the Helmand province in southern Afghanistan. Their mission is to locate Taliban leaders believed to be hiding there

ABOVE On June 2, 2003, as Phase III of Operation Mountain Lion continues, a cave has been set on fire as step one of cave destruction. The soldiers from B Company, 2nd Battalion, 187th Infantry Regiment, 101st Airborne Division (Air Assault) cleared it to ensure there are no Al Qaeda or Taliban inside. The cave pictured here is located near the Pakistan border. Clearing caves is a small part of sensitive site exploitation (SSE). (U.S. Army)

LEFT A member of the Afghanistan Military Forces (AMF) has a cigarette while on guard. The AMF are responsible for keeping the locals away while coalition forces headed by the U.S. Army Criminal Investigation Division (CID) unearth graves in the village of Markhanai on May 5, 2002 in Tora Bora, for Operation Torii. (USAF)

February 19–20, 2003 About 200 U.S troops from the 82nd Airborne Division are ferried by helicopters into the Sami Ghar mountains, about 60 miles (100km) east of Kandahar, initiating Operation Valiant Strike. The objective is to locate Osama Bin Laden and members of Al Qaeda

August 20, 2003 In a new offensive dubbed Operation Mountain Viper, the U.S. Army and the Afghan National Army (nearly 1,000 in number) work together in late August and early September, 2003, to uncover hundreds of suspected Taliban rebels dug into the mountains of Daychopan district, Zabul province, Afghanistan

December 6, 2003 The U.S. military launches its biggest-ever ground operation, Operation Avalanche, across eastern and southern Afghanistan. Over 2,000 soldiers are involved, including four infantry battalions as well as soldiers from the Afghan National Army and militia

Sources:

http://www.dod.gov, Department of Defense, U.S.

http://lcweb2.loc.gov/frd/cs/afghanistan/afghanistan.html, U.S. Library of Congress.

http://en.wikipedia.org/wiki/Timeline_of_Afghan_history, Afghanistan Timeline.

Local Afghani girls go about their chores collecting water, while coalition forces headed by the U.S. Army Criminal Investigation Division (CID) unearth graves in their village of Markhanai (USAF)

Design and development

Caves: natural and manmade

One of the major problems the Soviet Union faced during its invasion into Afghanistan was the difficulty in expelling Mujahideen from the cave and tunnel complexes dug into the sides of mountains. Few reports or other written evidence are on hand to accurately describe the reality on the ground. Nonetheless, there are a handful of former CIA operatives familiar with the clandestine conduct of the war against the Soviets. Some of them have stated that few natural caves exist but that the majority of caves are manmade in such a manner as to avoid direct bombardment by either air or artillery. The purpose of the overwhelming majority of these manmade caves was, and still is, to hold caches of weapons and ammunition. One CIA officer states that the dugouts where Al Qaeda are thought to have hidden are about 10 to 30ft (3 to 9m) deep and that "you have to drop bombs down sheer walls and rock faces. Most were built to be safe from any air or missile attack."

There are several basic types of cave/tunnel systems within Fortress Afghanistan:

- The simple cliff overhangs and natural crevices — both as old as the mountains.
- The *karez* irrigation system and wells, developed by farmers for agricultural purposes.
- Natural caves further developed by farmers or Mujahideen.
- Caves that are completely manmade in construction.
- Complex cave and tunnel systems predominantly developed during the occupation of Soviet forces from 1979 to 1989, as encountered in Tora Bora and Zhawar Kili.

It should not be overlooked that some of the caves have been used throughout history for habitation, mineral extraction and for religious purposes. However, the cave and cache proliferation is a phenomenon that followed the Soviet invasion in 1979. During the current war, Coalition troops have searched hundreds of caves, spending most of their time disposing of ordnance and destroying cave entrances. The deepest cave thus far encountered is no more than 60ft deep. The multi-layered, complex cave systems able to accommodate trucks or even tanks as seen in the media graphics are mere speculation as none have actually been discovered thus far. One complex at Zhawar Kili in the hotly contested Paktia Province did prove an exception as Lester Grau points out:

The difficulties of patrolling mountainous regions can be made no clearer then in this picture as a weapons platoon with mortars attempts to scale this fortress-like area. (U.S. Army)

As the base expanded, Mujahideen used bulldozers and explosives to dig at least 11 major tunnels. Some of these huge tunnels reached 1,640ft (500m) and contained a hotel, a mosque, arms depots and repair shops, a garage, a medical point, a radio center and a kitchen. A gasoline generator provided power to the tunnels and the hotel's video player. This impressive base became a mandatory stop for visiting journalists, dignitaries and other "war tourists." Apparently, this construction effort also interfered with construction of fighting positions and field fortifications.

Manmade cave complex locations are not only due to natural positions but are also dependent on geopolitical factors. The most obvious would be neighboring Pakistan, a strong supporter of the Mujahideen movement and some say creator of the Taliban. As a natural consequence of their support and given the very nature that both countries occupy the same mountain ranges, more caches and caves were built in those regions during the Soviet occupation. Mujahideen could easily flee to Pakistan and return virtually unmolested. During the proliferation of tunnel construction, heavy equipment was brought over from Pakistan.

Geology

J. Stephen Schindler writes in his article *Afghanistan: Geology in a Troubled Land* that geological maps could delineate carbonate lithologies (rock compositions) expected to contain karst, a distinctive formation where underlying bedrock is

U.S. forces named the caves/complexes after family members. The cave complex containing these munitions was named "Roberts" and contains numerous ammunition crates. Caches as well developed as this one this are not as common as simpler caves. (DoD)

dissolved by surface or groundwater, leaving a cracked appearance with natural cave formations; but the number of citations in the literature for significantly large natural caves in Afghanistan is low. Because of the lack of detailed geological maps of Afghanistan, identifying the lithology at a specific tunnel can be difficult. Fortunately, remote satellite imagery, particularly hyperspectral imagery, can reveal valuable clues to the specific lithology of an arid region.

The tunnels at Tora Bora, which have become well known in recent months, are in a category all their own. The area surrounding Tora Bora is known as the Kohe Sofaid, or also as the Spinghar mountain range. The dominant lithology is metamorphic gneiss and schist, forms of stratified metamorphic rock. Tunnels that were initially developed during the Russian invasion by the Afghan Mujahideen, were apparently expanded in recent years by Al Qaeda. With his engineering background and financing, Osama Bin Laden used hard-rock mining techniques to expand and enhance the large tunnels. Many smaller tunnels in the area were also developed using less sophisticated techniques and appear to be preferentially dug in softer rocks, such as schist, and others that are highly fractured.

Complete destruction of the larger tunnels will be difficult if they are located in gneiss or other crystalline rocks. Collapsing the entrance may be the most that can be accomplished in these more massive rocks. However, a tunnel in softer rock, such as a clastic sediment, may be more easily destroyed.

In a *New York Times* article republished by the *Tulsa World* on November 26, 2001, Dr. John F. Shroder, a professor of geology at the University of Nebraska and expert on Afghanistan's complex geology, has stipulated that natural caves do exist and some are miles long. Limestone, as well as crystalline rock made of minerals like quartz and feldspar, are found throughout Afghanistan's mountains. The latter is twice as hard as limestone and similar to those of Cheyenne Mountain, Colorado, where the North American Aerospace Defense Command, better known as NORAD, has its communications center deep underground. "This kind of rock is extremely resistant and it's a good place to build bunkers, and Bin Laden knows that."

Nevertheless, both men agree that the manmade caves can be particularly difficult to destroy permanently. The Soviets in one instance bombed the Zhawar Kili area, destroying an entrance to a cave hiding a hundred or so Mujahideen, only to have another bomb blast the entrance open. Recent combat and reconnaissance patrols in the same area have shown that Coalition forces may be able to destroy some of the entrances but not shut down the cave permanently.

Soldiers from the Canadian Princess Patricia's Light Infantry enter the village of Markhanai on May 5, 2002, in Tora Bora. (USAF)

Weapons systems at war

United States forces have used a variety of high-technology weapons to root out enemy fighters burrowed within the mountain ranges of eastern Afghanistan. Satellite imaging, unmanned drones and an armada of multiplatform aircraft provide complete dominance of the skies. Secretary of Defense Donald Rumsfeld and Vice President Dick Cheney are both long-term advocates of technology-driven wars fought with few "boots on the ground." Much to the consternation of senior military officers, this strategy ultimately cost the Coalition their deepest desire: Osama Bin Laden. As the war on Al Qaeda started, the aerial campaign should have incorporated a massive ground assault by Coalition and Pakistani forces in the Tora Bora region where Bin Laden and hundreds of his devoted followers had retreated to in order to fight from their well-constructed caves. There are reports that Bin Laden paid local farmers cash and provided them with drills to enhance existing caves or to dig new ones prior to the American attack. Bombing runs complemented with a handful of special operations forces, and anti-Taliban fighters with varied commitment and dedication made for a shortsighted and unsuccessful strike. Although dozens of Taliban/Al Qaeda (TAQ) fighters were killed, ultimately several hundred escaped to confront Coalition forces with resounding success a few months later during Operation Anaconda. Subsequently, a number of dead fighters were exhumed from a burial site in Tora Bora by Coalition forces in order to retrieve DNA samples. Locals told journalist Philip Shmucker that hundreds of Al Qaeda supporters with many vehicles

ABOVE Airman 1st Class James Blair coordinates air cover for Army 10th Mountain Division light-infantry soldiers during recent operations in the Sroghar Mountains. Blair and other tactical air control party airmen are serving with Special Operations Forces (SOF) in operations Enduring and Iraqi Freedom. (USAF)

RIGHT Destroyed artillery. Quite possibly the result of an intensive American bombing run in eastern Afghanistan. (U.S. Army)

had escaped to Tora Bora. Coalition claims that large numbers of fighters and vehicles were destroyed were disputed by the locals, who claimed that hundreds were guided to safety in Pakistan.

The U.S.-led Coalition, by its misguided emphasis on waging primarily a technology-driven campaign, is directly to blame for the dispersal of key TAQ individuals. Although a number of key enemy personnel have been killed or captured subsequently, the initial American failures at Tora Bora and during Operation Anaconda have further enhanced the reputation of the TAQ militias and Osama Bin Laden, Al Qaeda's key public figure and financier.

Nonetheless, the arsenal of the United States is impressive, and, along with its industrial/military/petroleum complex, has created history's most powerful global empire.

During the early months of Operation Enduring Freedom alone, 70 percent of all munitions used by the air component were precision-guided and more than 85,000 sorties were flown, dropping more than 30,750 munitions weighing more than 9,650 tons. The cost effectiveness of the bombing campaign vis-à-vis actual accomplishments, however, is disproportionate. Although arguments can be made that the Taliban self-elected government was destroyed and the Al Qaeda organization scattered, and thus military victory achieved, the initiative was lost. Three years into Operation Enduring Freedom, Afghanistan is less safe, and Taliban, Al Qaeda, foreign fighters and disgruntled Afghan warlords dominate the country, particularly the eastern mountain ranges. The failure to apprehend Al Qaeda's key personnel at the outset rests squarely on the American architects of the war, Secretary of Defense Rumsfeld and Assistant Secretary of Defense Wolfowitz. Both men created faulty scenarios for easy victories in Afghanistan and Iraq, arguing that no one could withstand the technologically advanced war machinery of the United States. The U.S. military's top brass, reduced to cheerleading and public relations, bears equal fault. The early attacks on the TAQ militias were akin to hammer blows on anthills and their subsequent failures have forced the senior leadership to reevaluate its strategic concept and recognize that placing more troops on the ground and reducing the reliance on technology, as the mountains and tunnel systems clearly interfere with communication devices and subsequent intelligence gathering. The most crucial aspect of winning a war against guerrillas in a hostile land is human intelligence, combined with soldiers on the ground patrolling aggressively.

There can be no doubt that the U.S. has the finest equipment in the world proportionate to its extravagant defense budget. Combined-arms doctrine to integrate the numerous battle operating systems is challenging, but the U.S. certainly has set the standards for its execution. Unfortunately, terrain and that intangible element, the will of the people, are not cooperating. The following is but a sample of the devastating arsenal on hand.

An A-10 Thunderbolt II takes off from Bagram Airbase to provide close air support to the Army's 10th Mountain Division as they pursue Taliban and Al Qaeda forces during Operation Mountain Resolve. (USAF)

Aircraft

A-10A/OA-10A THUNDERBOLT II

The A/OA-10 Thunderbolt II is specifically designed to act in the role of close air support for forces on the ground. It is a simple design and can absorb a great deal of punishment. It also carries a wide range of different munitions, enabling it to engage a number of different targets including AFVs.
Power plant: Two General Electric TF34-GE-100 turbofans
Thrust: 9,065lb each engine
Length: 53ft 4in. (16.16m)
Height: 14ft 8in. (4.42m)
Wingspan: 57ft 6in. (17.42m)
Speed: 420mph (Mach 0.56)
Ceiling: 45,000ft (13,636m)
Maximum Takeoff Weight: 51,000lb (22,950kg)
Range: 800 miles (695 nautical miles)
Armament: One 30mm GAU-8/A seven-barrel Gatling gun; up to 16,000lb (7,200g) of mixed ordnance on eight under-wing and three under-fuselage pylon stations, including 500lb (225kg) of Mk-82 and 2,000lb (900kg) of Mk-84 series low/high drag bombs, incendiary cluster bombs, combined effects munitions, mine-dispensing munitions, AGM-65 Maverick missiles and laser-guided/electro-optically guided bombs; infrared countermeasure flares; electronic countermeasure chaff; jammer pods; 2.75in. (6.99cm) rockets; illumination flares and AIM-9 Sidewinder missiles.
Crew: One

ABOVE An Air Force weapons loader from the 28th Air Expeditionary Wing preps a 2,000lb bomb to be loaded into a B-1B Lancer bomber during Operation Enduring Freedom. Air Force B-2 Spirit, B-1B Lancer and B-52 Stratofortress aircraft have dropped more than 80 percent of the tonnage on combat missions over Afghanistan. (USAF)

AC-130H/U GUNSHIP

The AC-130 gunship is a heavily armed variant of the C-130 Hercules airframe. It is primarily used for close air support, air interdiction and force protection. Missions in close air support are troops in contact, convoy escort and urban operations. Air interdiction missions are conducted against preplanned targets or targets of opportunity. Force protection missions include air base defense and facilities defense.
Power plant: Four Allison T56-A-15 turboprop engines
Thrust: 4,910 shaft horsepower each engine
Length: 97ft 9in. (29.8m)
Height: 38ft 6in. (11.7m)
Wingspan: 132ft 7in. (40.4m)
Speed: 300mph (Mach 0.4) (at sea level)
Range: Approximately 1,300 nautical miles; unlimited with air refueling.
Ceiling: 25,000ft (7,576m)
Maximum Takeoff Weight: 155,000lb (69,750lb)
Armament: AC-130H—two 20mm M61A1 cannons, one 40mm L60 gun, one 105mm M102 howitzer; AC-130U—one 25mm GAU-12 cannon, one 40mm L60 gun, one 105mm M102 howitzer
Crew: Five officers (pilot, co-pilot, navigator, fire control officer, electronic warfare officer) and eight enlisted men (flight engineer, TV operator, infrared detection set operator, loadmaster, four aerial gunners)

B-1B LANCER

Carrying the largest payload of both guided and unguided weapons in the Air Force inventory, the multi-mission B-1

RIGHT Reminiscent of the Vietnam War, Canadian troops offload from a Chinook transport helicopter. This image provides a better understanding of the terrain and situation Coalition units encountered during Operation Anaconda. (DoD)

LEFT The mission of the MC-130E Combat Talon I and MC-130H Combat Talon II is to provide global, day, night and adverse weather capability to airdrop and air land personnel and equipment in support of U.S. and allied Special Operations Forces (SOF). The MC-130E also has a deep penetrating helicopter refueling role during special operations missions. These aircraft are equipped with in-flight refueling equipment, terrain-following, terrain-avoidance radar, an inertial and global positioning satellite navigation system, and a high-speed aerial delivery system. The special navigation and aerial delivery systems are used to locate small drop zones and deliver people or equipment with greater accuracy and at higher speeds than possible with a standard C-130. The aircraft is able to penetrate hostile airspace at low altitudes and crews are specially trained in night and adverse weather operations. (USAF)

LEFT Ground crew members wave at a B-52H Stratofortress bomber as it taxis for takeoff on a strike mission against Al Qaeda terrorist training camps and military installations of the Taliban regime in Afghanistan on October 7, 2001, during Operation Enduring Freedom. (USAF)

is the backbone of America's long-range bomber force. It can rapidly deliver massive quantities of precision and non-precision weapons against any adversary, anywhere in the world, at any time.

Eight B-1s were deployed in support of Operation Enduring Freedom. B-1s dropped nearly 40 percent of the total tonnage during the first six months of OEF. This included nearly 3,900 JDAMs, or 67 percent of the total.

Power plant: Four General Electric F-101-GE-102 turbofan engines with afterburner
Thrust: over 30,000lb with afterburner, per engine
Length: 146ft (44.5m)
Wingspan: 137ft (41.8m) extended forward, 79ft (24.1m) swept aft
Height: 34ft (10.4m)
Weight: Empty, approximately 190,000lb (86,183kg)
Maximum Takeoff Weight: 477,000lb (216,634 kg)
Speed: 900-plus mph (Mach 1.2 at sea level)
Range: Intercontinental, unrefueled
Ceiling: More than 30,000ft (9,144m)
Crew: Four (aircraft commander, copilot, and two weapon systems officers)
Armament: 24 GBU-31 GPS-aided JDAM (both Mk-84 general purpose bombs and BLU-109 penetrating bombs) or 24 Mk-84 2,000lb general purpose bombs; 84 Mk-82 500lb general purpose bombs; 84 Mk-62 500lb naval mines; 30 CBU-87, -89, -97 cluster munitions.

B-2A SPIRIT

The B-2A is a multi-role bomber capable of delivering both conventional and nuclear munitions. A dramatic leap forward in technology, the bomber represents a major milestone in the U.S. bomber modernization program. The B-2 brings massive firepower to bear, in a short time, anywhere on the globe through previously impenetrable defenses.

Power Plant: Four General Electric F-118-GE-100 engines
Thrust: 17,300lb each engine
Length: 69ft (20.9m)
Height: 17ft (5.1m)
Wingspan: 172ft (52.12m)
Speed: High subsonic
Ceiling: 50,000ft (15,240m)
Takeoff Weight (Typical): 336,500lb (152,634kg)

Range: Intercontinental, unrefueled
Armament: Conventional or nuclear weapons
Payload: 40,000lb (18,144kg)
Crew: Two pilots

B-52H STRATOFORTRESS

Air Combat Command's B-52H is a long-range, heavy bomber that can perform a variety of missions. The bomber is capable of flying at high subsonic speeds at altitudes up to 50,000ft (15,240m). It can carry nuclear or precision-guided conventional ordnance with worldwide precision navigation capability.
Power plant: Eight Pratt & Whitney engines TF33-P-3/103 turbofan
Thrust: Each engine up to 17,000lb
Length: 159ft 4in. (48.5m)
Height: 40ft 8in. (12.4m)
Wingspan: 185ft (56.4m)
Speed: 650 miles per hour (Mach 0.86)
Ceiling: 50,000ft (15,151.5m)
Weight: Approximately 185,000lb empty (83,250kg)
Range: Unrefueled 8,800 miles (7,652 nautical miles)
Armament: Approximately 70,000lb (31,500kg) mixed ordnance—bombs, mines and missiles. (Modified to carry air-launched cruise missiles, Harpoon anti-ship and Have Nap missiles.) One 20mm tail gun.
Crew: Five (aircraft commander, pilot, radar navigator, navigator and electronic warfare officer

EC-130E/J COMMANDO SOLO

The EC-130E/J Commando Solo, a specially modified four-engine Hercules transport, conducts information operations, psychological operations and civil affairs broadcasts in AM, FM, HF, TV and military communications bands. A typical mission consists of a single-ship orbit offset from the desired target audience—either military or civilian personnel. Most recently, the aircraft broadcast messages to the local Afghan population and Taliban soldiers during Operation Enduring Freedom.

EC-130H COMPASS CALL

Compass Call is the designation of a modified version of the C-130 Hercules aircraft configured to perform tactical information warfare. Specifically, the modified aircraft prevents or degrades communications essential to command and control of weapon systems and other resources. The system primarily supports tactical air operations, but also can provide jamming support to ground forces and amphibious operations.

F-15E STRIKE EAGLE

The F-15E is a dual-role fighter designed to perform air-to-air and air-to-ground missions. An array of avionics and electronics systems gives the F-15E the capability to fight at low altitude, day or night, and in inclement weather and perform its primary function as an air-to-ground attack aircraft.
Power plant: Two Pratt & Whitney F100-PW-220 or 229 turbofan engines with afterburners
Thrust: 25,000–29,000lb each engine
Wingspan: 42.8ft (13m)
Length: 63.8ft (19.44m)
Height: 18.5ft (5.6m)
Speed: Mach 2.5 plus
Maximum takeoff weight: 81,000lb (36,450kg)
Service ceiling: 50,000ft (15,240m)
Range: 2,400 miles (3,840km) ferry range with conformal fuel tanks and three external fuel tanks
Armament: One 20mm M61A1 six-barrel cannon mounted internally with 500 rounds of ammunition. Four AIM-7F/M Sparrow missiles and four AIM-9L/M Sidewinder missiles, or eight AIM-120 AMRAAM missiles. Any air-to-surface weapon in the Air Force inventory (nuclear and conventional)
Crew: Pilot and weapon systems officer

F-16C/D FIGHTING FALCON

The F-16 is a highly flexible, maneuverable multi-role fighter aircraft. Its versatility is such that it is equally suited to air-to-air and air-to ground combat. It has proved a popular airframe both with U.S. forces and allied forces, as it has been widely exported.
Power plant: F-16C/D: one Pratt and Whitney F100-PW-200/220/229 or General Electric F110-GE-100/129
Thrust: 27,000lb
Length: 49ft 5in. (14.8m)
Height: 16ft (4.8m)
Wingspan: 32ft 8in. (9.8m)
Speed: 1,500mph (Mach 2 at altitude)
Ceiling: Above 50,000ft (15,240m)
Maximum Takeoff Weight: 37,500lb (16,875kg)
Range: More than 2,000 miles ferry range (1,740 nautical miles)
Armament: One 20mm M61A1 six-barrel cannon with 500 rounds; external stations can carry up to six air-to-air missiles, conventional air-to-air and air-to-surface munitions and electronic countermeasure pods
Crew: F-16C, one; F-16D, one or two

HH-60G PAVE HAWK

The primary mission of the HH-60G Pave Hawk helicopter is to conduct day or night operations into hostile environments to recover downed aircrew or other isolated personnel during war. Because of its versatility, the HH-60G is also tasked to perform military operations other than war. These tasks include civil search and rescue, emergency aeromedical evacuation, disaster relief, international aid, counterdrug activities and NASA space shuttle support.
Power plant: Two General Electric T700-GE-700 or T700-GE-701C engines
Thrust: 1,560–1,940 shaft horsepower, each engine

Length: 64ft 8in. (17.1m)
Height: 16ft 8in. (4.4m)
Rotor Diameter: 53ft, 7in. (14.1m)
Speed: 184mph (294.4kph)
Maximum Takeoff Weight: 22,000lb (9,900kg)
Range: 445 statute miles; 504 nautical miles (unlimited with air refueling)
Armament: Two 7.62mm machine guns
Crew: Two pilots, one flight engineer and one gunner

RQ-4A GLOBAL HAWK

The Global Hawk Unmanned Aerial Vehicle (UAV) provides Air Force and joint battlefield commanders near-real-time, high-resolution, intelligence, surveillance and reconnaissance imagery. In the last year, the Global Hawk provided Air Force and joint warfighting commanders more than 15,000 of these images to support Operation Enduring Freedom, flying more than 50 missions and 1,000 combat hours to date.
Wingspan: 116ft (35.3m)
Length: 44ft (13.4m)
Range: 13,800 statute miles; 12,000 nautical miles
Ceiling: 65,000ft (19,812m)
Speed: 340 knots (about 400mph)
Weight 25,600lb (11,612kg)

MH-53J/M PAVE LOW

The Pave Low's mission is low-level, long-range, undetected penetration into denied areas, day or night, in adverse weather, for infiltration, exfiltration and resupply of special operations forces.
Power Plant: Two General Electric T64-GE/-100 engines
Thrust: 4,330 shaft horsepower per engine
Length: 88ft (28m)
Height: 25ft (7.6m)
Rotary Diameter: 72ft (21.9m)
Speed: 165mph (at sea level)
Ceiling: 16,000ft (4,876m)
Maximum Takeoff Weight: 46,000lb
Range: 640 statute miles; 600 nautical miles (unlimited with aerial refueling)
Armament: Combination of three 7.62mm mini guns or three .50-cal. machine guns
Crew: Officers, two pilots; enlisted, two flight engineers and two aerial gunners

ABOVE An A-10 flies close to the ground under hostile circumstances and looses flares to throw off any surface-to-air missiles, September 1, 2003, during Operation Mountain Viper. (U.S. Army)

LEFT An F/A 18 Hornet from Strike Fighter Squadron Eight Two (VFA 82 "Marauders") soars through the sky in support of Operation Enduring Freedom, November 14, 2003. (USN)

RQ-/MQ-1 PREDATOR UNMANNED AERIAL VEHICLE

The RQ-1 and MQ-1 Predators are medium-altitude, long-endurance unmanned aerial vehicle systems. The MQ-1's primary mission is interdiction and conducting armed reconnaissance against critical, perishable targets. When the MQ-1 is not actively pursuing its primary mission, it augments the RQ-1 as a Joint Forces air component commander-owned theater asset for reconnaissance, surveillance and target acquisition in support of the Joint Force commander.

The RQ-1 and MQ-1 Predators are systems, not just aircraft. A fully operational system consists of four aircraft (with sensors), a ground control station (GCS), a Predator Primary Satellite Link, and approximately 82 personnel for continuous 24-hour operations.

The basic crew for the Predator is one pilot and two sensor operators. They fly the aircraft from inside the GCS via a C-Band line-of-sight data link or a Ku-Band satellite data link for beyond line-of-sight flight. The aircraft is equipped with a color nose camera (generally used by the aerial vehicle operator for flight control), a day variable aperture TV camera, a variable aperture infrared camera (for low light/night), and a synthetic aperture radar for looking through smoke, clouds or haze. The cameras produce full motion video and the SAR still frame radar images.

The MQ-1 Predator carries the Multispectral Targeting System with inherent AGM-114 Hellfire missile targeting capability and integrates electro-optical, infrared, laser designator and laser illuminator into a single sensor package. The aircraft can employ two laser-guided Hellfire anti-tank missiles with the MTS ball.

Power plant: Rotax 914 four-cylinder engine producing 101 horsepower
Length: 27ft (8.22m)
Height: 6.9ft (2.1m)
Weight: 1,130lb (512kg) empty, maximum takeoff weight 2,250lb (1,020kg)
Wingspan: 48ft 7in. (14.8m)
Speed: Cruise speed around 84mph (70 knots), up to 135mph
Range: up to 454 statute miles (400 nautical miles)
Ceiling: up to 25,000ft (7,620m)

Munitions

JOINT DIRECT ATTACK MUNITIONS GBU 31/32

The Joint Direct Attack Munition (JDAM) is a guidance tail kit that converts existing unguided free-fall bombs into accurate, adverse weather "smart" munitions. With the addition of a new tail section that contains an inertial navigational system and a global positioning system guidance control unit, JDAM improves the accuracy of unguided, general purpose bombs in any weather condition. JDAM is a joint U.S. Air Force and Department of Navy program.

JDAM enables multiple weapons to be directed against single or multiple targets on a single pass. JDAM is currently compatible with B-1B, B-2A, B-52H, F-15E, F-14A/B/D, F/A-18E/F, F-16C/D and F/A-18C/D aircraft.

Primary Function: Guided air-to-surface weapon
Length: (JDAM and warhead) GBU-31 (v) 1/B: 152.7in. (387.9cm); GBU-31 (v) 3/B: 148.6in. (377.4cm); GBU-32 (v) 1/B: 119.5in. (303.5cm)
Launch Weight: (JDAM and warhead) GBU-31 (v) 1/B: 2,036lb (925.4kg); GBU-31 (v) 3/B: 2,115lb (961.4kg); GBU-32 (v) 1/B: 1,013lb (460.5kg)
Wingspan: GBU-31: 25in. (63.5cm); GBU-32: 19.6in. (49.8cm)
Range: Up to 15 miles
Ceiling: over 45,000ft (13,677m)

RQ-1 Predator unmanned aerial vehicles, like this one, have been used to increase battlefield awareness at operating locations in support of Operation Enduring Freedom. Intelligence, surveillance and reconnaissance (ISR) assets, including unmanned aerial vehicles, RC-135V/W Rivet Joint, U-2A Dragon Lady, E-3A Sentry Airborne Warning and Control System (AWACS), and EC-130E/H Commando Solo, have flown more than 325 missions to provide battlefield awareness in support of OEF. (USAF)

THERMOBARIC WEAPONS

The word "thermobaric" has roots in the ancient Greek ("therme"—"heat"—and "baros"—pressure) and derives from the combined effect of temperature and pressure brought upon the target. Thermobaric weapons were first developed during the Vietnam War by the U.S. in the form of fuel-air explosives, used with devastating effect. The primary thermobaric bomb, designated as BLU-118/B (Bomb Live Unit), was developed to support combat operations in Operation Enduring Freedom. On December 21, 2001, the Department of Defense officially announced that a small number of the weapons were being deployed to attack tunnels in Afghanistan. No official reports are available to assess their effectiveness, though certainly they are extravagantly cruel weapons. On March 3, 2002, a single 2,000lb thermobaric bomb was used for the first time in combat against cave complexes in which Al Qaeda and Taliban fighters had taken refuge in the Gardez region of Afghanistan during Operation Anaconda. Thermobaric bombs were also employed during Operation Iraqi Freedom.

Explosion of a JDAM air strike on a target identified by SOF recon teams. (Tactical Assault Gear)

A small crater, created by a 500lb Joint Direct Attack Munition on March 23, 2004, near a destroyed compound that housed suspected anti-Coalition forces in Uruzgan Province along the Tirin River. (U.S. Army)

Offensive operations

Although the Taliban and Al Qaeda militias were easily defeated, their leadership was able to reconstitute their forces and over the last two years have regained much of their lost territories. The Coalition forces have increased their numbers and training for a fledgling Afghan national army is ongoing. The inevitability of fighting a protracted guerrilla war in Afghanistan's mountain ranges leads to the natural conclusion that high technology and airpower alone do not suffice for a successful campaign. After all, age-old maxims of war still hold true in the 21st century: infantry needs to hold the ground, even if only temporarily. The advantage in mountainous terrain lies with the defender and overwhelming force needs to be brought to bear on the enemy's weak points or centers of gravity. The most recent combat operations of foreign armies against Mujahideen fighters from 1979 to 2004 have painted a campaign of small groups of men with small arms battling against enormous odds of technology-driven superpowers. Most caves are commonly used as caches for food and weapons and, at times, as central command facilities for local militia units.They also served as quarters, medical aid stations, shelters against air attack, and hideouts for couriers.

How does one "win" against caves, fortified houses and tunnel systems that serve indigenous fighters in many ways? How does one beat the militias whose caves serve as hiding places, protection against observation, air strikes and other incoming artillery and small-arms fire?

Soviet procedures against *karez* systems

In general terms Soviet forces tended to shy away from flushing out hidden Mujahideen unless it was part of an overall operation, and at that point special attention was paid to caves/tunnels. The Soviet Army taught special courses on tunnel neutralization for communist Afghan government forces akin to a program designed for Soviet sappers in Afghanistan. Senior Soviet military leadership realized a need for such difficult tasks, as soldiers are hesitant to enter dark underground places. Additionally, the intrinsic nature of this type of fighting was fraught with numerous problems. Proximity to enemy combatants, booby traps, manmade and natural ones such as scorpions or snakes.

Karez systems were easy to find and the standard operating procedure was for two groups to secure two adjacent shafts. After a warning shout, or not as the case

Karez water system. This aerial shot shows the aboveground wells from which farmers draw water for irrigation. The area seems to be south of Kandahar. (DoD)

may be, two concussion hand grenades were thrown in. At this point a rock would be tossed down the shaft to determine the depth, as that information would be needed to prepare charges for the destruction of the tunnel. A mirror reflecting light into the shaft would help to examine it and blind spots would be addressed by lowering another grenade attached to a cord to that particular area.

September 29, 2003, soldiers of 1st Platoon, C Company, 1st Battalion 87th Infantry Regiment (+), stationed at Orgun-E firebase, Afghanistan, conduct Operation Tri-City Sweep to conduct a cordon and sweep of Say Khan, Gulmani Kot and Pir Kowti, three towns that lie within the Pir Kowti Valley. The soldiers are looking for weapon caches and weapons deemed illegal to own as well as any evidence of Taliban activity. (U.S. Army)

> Only after this, would the soldiers lower a blasting charge on detonating cord. Usually they would use captured TS-2.5 or TS-6.1 Italian anti-vehicular blast mines. They had many of these available. The soldiers would lower the first charge to the bottom of the shaft. Then they would prepare a second charge using three or four meters of detonating cord and 800g ($2^3/_4$lb) of high explosive. Then they would tie or tape the detonation cord to a standard hand grenade fuse. They would weight down this firing assembly with a rock or wedge it near the shaft mouth. Two trained soldiers could prepare a 20m shaft for detonation in about three minutes. All that remained was to pull the ring on the firing assembly. After four seconds, the charges would explode. During the explosion, it was necessary to stand some 5 or 6m from the mouth of the shaft, since the explosion would throw rocks out like a volcano.
> Lester W. Grau and Ali Ahmed Jallai, *Underground Combat: Stereophonic Blasting, Tunnel Rats and the Soviet-Afghan War*

This charge placement was particularly effective since the top charge would explode a fraction of a second earlier than the bottom charge. This top explosion would tightly plug the shaft with gases. Then the bottom charge would explode. The shock wave from this explosion would rebound off the higher gas mass and rush back down and against the sides of the shaft and tunnels. This creates a deadly over-pressure between the two charges, which the Soviets called "the stereophonic effect." The stereophonic effect could be amplified with the addition of a second shaft being blown simultaneously by joining the separate charges to a single detonation cord, called the "quadronic effect."

Smoke pots would then be dropped into the shafts and if the smoke disappeared the tunnels would still be intact and a team of three to four men could enter without protective masks. The pointman would have his leg tied to a line. This had several purposes. He could be dragged up if wounded or killed, or it could be used to haul other equipment or material found in the location. The tunnel rats were armed with knifes, entrenching tools, hand grenades, pistols and assault rifles. Some of the assault rifles had flashlights attached to the forestock. Tracer rounds were used to fill up the magazines.

Another weapon used by the Soviets in clearing tunnels was the SM signal projector, a sort of Roman candle with a psychological impact. Designed for trip-wire release by an unwary enemy, the SM signal emitted a siren-like sound and launched red, green or white signal stars up to 20m. Up to six of these candles could be taped together and ignited to the front of the search team. The flame would shoot out for

> nine seconds, a brilliant shaft of light, screams of sirens, and a fountain of signal stars would fill the tunnel. The signal stars would ricochet off the tunnel walls like tracers. The Soviets would find the unsuspecting foe covering his head with his arms, even though there was no real danger unless a signal star hit someone in the eye.
> Lester W. Grau and Ali Ahmed Jallai, *Underground Combat: Stereophonic Blasting, Tunnel Rats and the Soviet-Afghan War*

Russian stereophonic warfare
Specially trained Russian soldiers would lower two charges into a *karez* irrigation system where, historically, the local population would seek refuge during times of war. The length of the ropes to which the charges were attached was determined by tossing rocks down the shaft. The top charge would explode first, thus sealing the top of the tunnel entrance, allowing for the secondary explosion to travel up, then back down against the shaft and any interior openings. This method of sensitive site exploration was preferable over sending men down the shafts and into the tunnel system. Undoubtedly, this kind of tunnel clearing was sophisticated but certainly it made very little difference in the overall campaign to kill the Mujahideen and win the "hearts and minds" of the indigenous population.

ABOVE LEFT Staff Sgt. Raybon Smith, the Squad Leader for 2nd Squad, 1st Platoon, A Company, 41st Engineer Battalion, uses the flashlight attached to his M4 carbine to light the cave so he and another soldier can set up a third 40lb shaped charge in order to destroy the cave on September 24, 2003. (U.S. Army)

ABOVE RIGHT On April 16, 2003, soldiers assigned to the 307th Engineer Battalion search for weapons that may be hidden in a well in the town of Khar Bolaq while participating in a mission called Operation Crackdown. (U.S. Army)

LEFT Soldiers of 2nd Platoon, C Company, 1st Battalion, 87th Infantry Regiment, conduct a raid in the village of Shaykhan looking for two Taliban members who were bragging about their part in the killing of American soldiers. Sergeant Wayne Trimble waits for Private Second Class Scott Edwards to enter a dwelling so they can search for weapons caches. (U.S. Army)

The Soviets also used different flamethrowers throughout the war. The RPO-A (Recoilless Infantry Flame Projector), also known as *shmel* ("bumblebee"), was the most widely distributed thermobaric weapon amongst Soviet troops. It is a single-shot, disposable, lightweight, shoulder-fired, recoilless rocket launcher.

> Three types of projectiles were used: thermobaric (fuel-air), incendiary and smoke. The fuel-air round was most effective against *karez*. The problem was that flamethrower gunners drew more small-arms fire than radiomen. An incendiary round from an RPO-A could clear out any opposition on the surface around a shaft entrance, but no flamethrower gunner wanted to lean over the mouth of a karez to fire down the shaft. He might be shot before he could get off a round. The Soviets would secure the shaft entrance and then lock and cock an RPO-A with a thermobaric round. They would tie two lowering lines on the RPO-A and a string on the trigger. Then they would slowly lower the RPO-A down the shaft until it was facing a tunnel. They would then pull the trigger string to fire the thermobaric round down the tunnel. The resulting over-pressure of the fuel-air round could be devastating.
> Lester W. Grau and Ali Ahmed Jallai, *Underground Combat: Stereophonic Blasting, Tunnel Rats and the Soviet-Afghan War*

The Soviets, no matter how ingenious or dedicated in their fighting, failed.

Fire team of the 101st Airborne Division in eastern Afghanistan. (U.S. Army)

U.S. procedures in tunnel operations

Although reports tend to exaggerate the complexities of some cave systems, there is no doubt that some are truly well developed, extensive complexes containing numerous areas and interconnecting tunnels. Others may be simpler, but any invading foe will need to deal with all of them in a careful and well-executed manner to avoid any potential disaster. Whether a cave system, reinforced stronghold or simple fighting-hole, these structures present a deadly obstacle to the men conducting combat or reconnaissance patrols.

Although the theory of combined arms has a longstanding tradition of bringing all assets to bear on the enemy, unconventional warfare and terrain-dictating campaigns may impede proper deployment. In any event, tackling caves, tunnels, trenchlines and strongholds in mountains requires one thing – men, climbing, "humping," fighting and bleeding. Mujahideen have done it successfully for generations. Coalition forces have not. Terrain and weather excise a heavy price on soldiers and equipment. Studies have shown not only the wear on troops and their gear, but simply put, modern soldiers can fight in high altitude for only a limited time before needing to refit and rest. Mujahideen can simply withdraw to safer areas, live with their families and prepare for the next raid or attack on the next target of opportunity.

Since the commencement of Operation Enduring Freedom, the U.S. Army Training and Doctrine Command has revisited older training manuals dealing with tunnel systems encountered in Vietnam and updated their handbook specifically for combat operations in Afghanistan. Handbook *02-8* describes in minute detail the tactics, techniques and procedures in operations against tunnels and cave complexes.

Tunnel characteristics

The first characteristic of a tunnel complex is normally superb camouflage. Entrances and exits are concealed, bunkers are camouflaged, and, even within the tunnel complex itself, side tunnels are concealed, both hidden trapdoors and dead-end tunnels are used to confuse the attacker. Usually, the first indication of a tunnel complex will be hostile fire from a concealed bunker, which might otherwise have gone undetected. Spoil from the tunnel system is normally distributed over a wide area, thus making it even harder to detect a cave/tunnel complex.

Trapdoors may be used, both at entrances and exits and inside the tunnel complex itself, concealing side tunnels and intermediate sections of a main tunnel. Trapdoors are of several types: They may be concrete covered by dirt, hard-packed dirt reinforced by wire, or a "basin" type consisting of a frame filled with dirt. This latter type is particularly difficult to locate as probing will not reveal the presence of the trapdoor unless a probe strikes the outer frame. Trapdoors covering entrances/exits are generally a minimum of 330ft (100m) apart. Booby traps may be used extensively, both inside and outside entrance/exit trapdoors.

Two soldiers from the 101st Airborne Division demonstrating the age-old "buddy system" specifically taught to World War II U.S. Rangers by their British Commando counterparts. These individual movement techniques are an integral part of small-unit tactics. (U.S. Army)

Recognition of their cellular nature is important for understanding tunnel complexes. Prisoner interrogation has indicated that many tunnel complexes are interconnected, but the connecting tunnels, concealed by trapdoors or blocked by 3 to 4ft of dirt, are known only to selected persons and are used only in emergencies. Indications also point to interconnections of some length, e.g., 3–5 miles, through which relatively large bodies of men may be transferred from one area to another, especially from one "fighting" complex to another. The "fighting" complexes terminate in well-constructed bunkers, in many cases covering likely landing zones in a war zone or base area.

Tunnel techniques

A trained tunnel exploitation and denial team is essential. Untrained personnel may miss hidden tunnel entrances and caches, take unnecessary casualties from concealed mines and booby traps, and may not adequately destroy the tunnel.

Tunnel teams should be trained, equipped, and maintained in a ready status to quickly explore a complex when discovered.

Careful mapping of a tunnel complex may reveal other hidden entrances as well as the location of adjacent tunnel complexes and underground defensive systems.

Small-caliber pistols or pistols with silencers are the weapons of choice in tunnels, since large-caliber weapons without silencers may collapse sections of the tunnel when fired and/or damage eardrums.

Personnel exploring large tunnel complexes should carry a colored smoke grenade to mark the location of additional entrances as they are found. In mountainous desert areas it is often difficult to locate the position of these entrances without smoke.

Two- and three-man teams should enter tunnels for mutual support.

Claustrophobia and panic could well cause the failure of the team's mission or the death of its members. This is why repetitive training is crucial.

Constant communication between the tunnel and the surface is essential to facilitate tunnel mapping and exploitation.

A representative equipment list for a tunnel team:

Protective mask – one per individual
TA-1 telephone – two each
One half-mile field wire on doughnut roll
Compass – two each
Sealed beam 12-volt flashlight – two each
Small-caliber pistol – two each
Probing rods – 12in. and 36in.
Bayonet – two each
Mity Mite Portable Blower – one each
M7A2 CS grenades – twelve each
Powdered CS-1 – as required
Colored smoke grenades – four each
Insect repellant and spray – four cans
Entrenching tool – two each
Cargo packs on pack board – three each
Source: U.S. Army Handbook 02-8

Tunnel exploitation and destruction

The area in the immediate vicinity of the tunnels is secured and defended by a 360-degree perimeter to protect the tunnel team. The entrance to the tunnel is carefully examined for mines and booby traps.

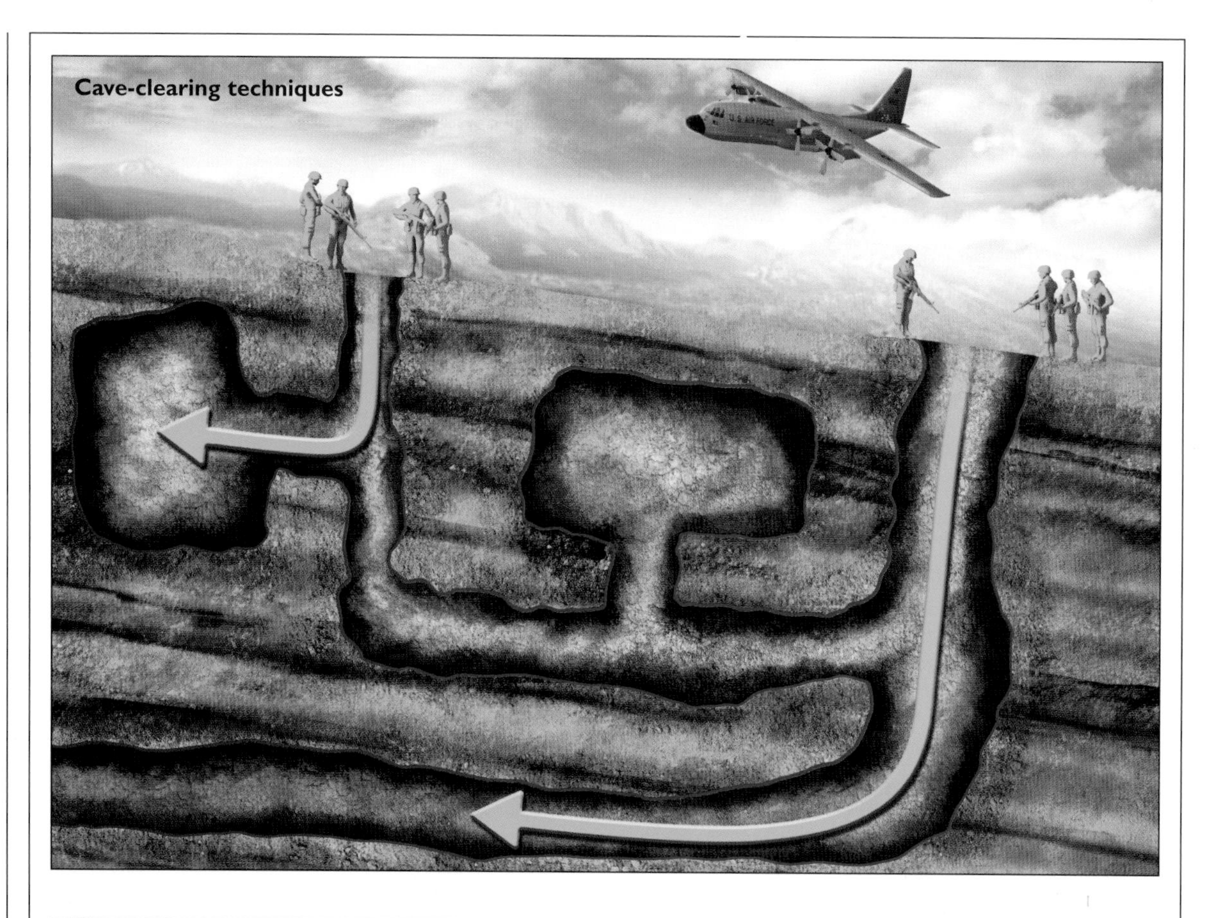

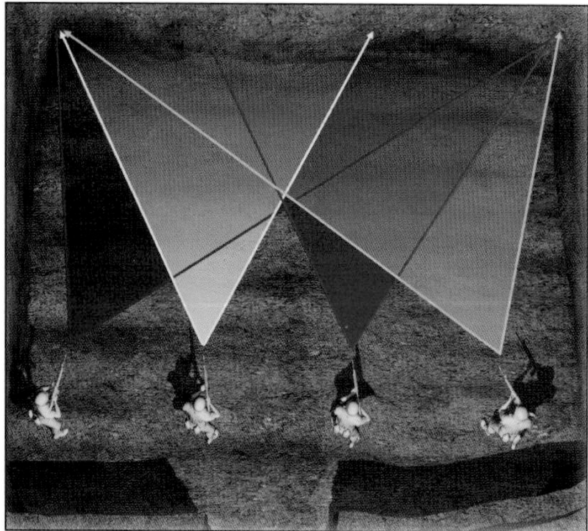

This diagram shows a four-man team clearing a room or a cave. Each different colored arc shows each man's area of responsibility. There are numerous techniques taught for clearing a room/cave but the principal element requires each soldier to work within the team and its standard operating procedure. Each member must be confident that the other will do his job by covering the appropriate sector.

This diagram shows a fire team moving through a tunnel in a Serpentine formation — three men to the front with the fourth man pulling rear security. There are other ways of moving such as the Diamond formation—a lead man, two flankers and rear security.

Cave-clearing techniques

Although few large complex cave/tunnel complexes exist, the fact still remains'that the infantry must clear or destroy such obstacles when encountered. Tunnel clearing, no matter the size of the actual complex, requires specialized training, as it is both arduous and nerve-racking. It requires confidence in training, equipment and teammates. Aerial and ground support secure the surrounding areas in order to shut down any potential escape routes and to provide support in case of enemy reaction forces moving toward the complex. Although similarities exist to modern-day room clearing techniques and close-quarters battle/combat, in tunnel clearing, grenades should not be used as the noise is deafening and may cause tunnels to collapse. An additional tool for the modern soldier in underground work is the sound suppressor designed to muffle the discharge of the assault rifle. Modern sensory devices and robots designed for this task are helpful, but no matter the technological tools and support available, the simple fact remains that a "footslogger" with a weapon has to go underground.

Two members of the team enter the tunnel with wire communications to the surface. The team works its way through the tunnel, probing with bayonets for booby traps and mines and looking for hidden entrances, food and arms caches, water locks, and air vents. As the team moves through the tunnel, compass headings and distances traversed are called to the surface. A team member at the surface maps the tunnel as exploitation progresses.

Captured arms and intelligence documents are secured and retrieved for destruction or analysis. Site exploitations have yielded important intelligence over the years.

Upon completion of exploitation, explosives are placed at all known tunnel entrances in order to seal each and prevent reuse by the enemy. If time or materials are not available for immediate closure, CS-1 Riot Control Agent can be placed at intervals down the tunnel at sharp turns and intersections. It must be emphasized, however, that the denial achieved by the use of CS-1 is only temporary in duration and used until demolitions are available to completely destroy the complex.

LEFT Two men clearing a cave entrance. One man is throwing a device used to trigger any potential booby trap wires. (U.S. Army)

Search and destroy operations must provide adequate time for a thorough search of the area to locate all tunnels. Complete exploitation and destruction of tunnel complexes is very time consuming, and operational plans must be made accordingly to ensure success.

The presence of a tunnel complex within or near an area of operations poses a continuing threat to all personnel in the area. No area containing tunnel

ABOVE LEFT Team clearing a long tunnel, throwing a three-pronged device to trigger any booby traps. Cave-clearing techniques must be executed properly and require a great deal of trust in your teammates. Even though Coalition forces regularly use airpower to destroy caves, "boots on the ground" are required to assess the damage and root out any intact enemy force. (U.S. Army)

ABOVE RIGHT Two men guarding a well-constructed cave entrance in eastern Afghanistan. (U.S. Army)

RIGHT One man staring down a long and dark winding tunnel. Special training in cave clearing is helpful to combat the natural fear factor of entering dark and potentially lethal areas. (U.S. Army)

complexes should ever be considered completely cleared. Too often small groups of enemy fighters engage Coalition forces in an area that had previously been thought to be cleared.

Tunnel flushing and denial

Time constraints or the threat of enemy action can force a tunnel team to use the Mity Mite Portable Blower (RVN, MACV 1965) to flush the enemy from tunnels. The Mity Mite can be used in conjunction with burning type CS Riot Control Agent grenades (M7A2), which have the additional effect or producing smoke which in most cases helps identify hidden entrances and air vents.

After flushing an entrance with CS grenades, the Mity Mite can then blow powdered CS-1 into tunnel entrances, rendering the tunnel unusable to the enemy for a short period of time, at least as far as the first "firewall" within the tunnel system.

Dangers

The dangers inherent to tunnel operations fall generally into the following categories and should be guarded against by all personnel connected with these operations:

- Presence of mines and booby traps in the entrance/exit area.
- Presence of small but dangerous concentrations of carbon monoxide produced by burning-type smoke grenades after tunnels are smoked. Protective masks will prevent inhalation of smoke particles, which are

One "grunt" crawling out of a bomb-ruined entrance. (U.S. Army)

On April 16, 2003, Sgt. Derrick Edwards (left) and PFC Brandon Chattman (right), assigned to the 307th Engineer Battalion, search for weapons that may be buried under a floor in a house in the town of Khar Bolaq while participating in a mission called Operation Crackdown. (U.S. Army)

dangerous only in very high concentration, but will not protect against carbon monoxide.

- Possible shortage of oxygen as in any confined or poorly ventilated space.
- Enemy still in the tunnel who pose a danger to friendly personnel both above and below ground (in some instances, dogs can successfully detect enemy hiding in tunnels).

A press release by the U.S. Army Special Operations Command reveals the immediacy and ferociousness of close-quarters combat (CQC) that can quickly degenerate into hand-to-hand combat. This rather insignificant action is a good example of adapting to a mission when aerial attacks are thought unsuitable.

Master Sgt. Anthony S. Pryor, a team sergeant with Company A, 1st Battalion, 5th SFG (Abn.), received the Silver Star Medal for his gallantry in

Another cleared entrance – well built. (U.S. Army)

Smoke drifts from a destroyed building possibly in the Shah-I-Kot valley. A foot patrol from the 101st may have been responsible for its destruction. (U.S. Army)

combat during the raid when he single-handedly eliminated four enemy soldiers, including one in unarmed combat, all while under intense automatic weapons fire and with a crippling injury.

On January 23, 2002, Pryor's company received an order from the U.S. Central Command to conduct their fourth combat mission of the war—a sensitive site exploitation of two compounds suspected of harboring Taliban and Al Qaeda terrorists in the mountains of Afghanistan. Because of the presence of women and children within the compounds, Pryor said aerial bombardment was not considered an option. Once on the ground there, the company was to search for key leadership, communications equipment, maps and other intelligence.

Sgt. 1st Class Scott Neil was one of the team members there with Pryor that night at the second compound. A Special Forces weapons sergeant, he fought on Pryor's team as a cell leader and found himself momentarily pinned down by the sudden hail of bullets after the team's position was compromised. "After the initial burst of automatic weapons fire, we

LEFT Soldiers of the 2nd Battalion, 505th PIR, move to a weapon cache site in the mountains of Afghanistan, March 28, 2003. Soldiers assigned to this unit were given the mission of conducting searches for weapons caches in the mountains of Afghanistan near Kohe Sofi. (U.S. Army)

ABOVE LEFT A soldier from B Company, 2nd Battalion, 504th Parachute Infantry Regiment, climbs into a small hole to search for suspected Taliban and weapon caches in the Bahgran Valley, during Operation Viper on February 19, 2003. (U.S. Army)

ABOVE RIGHT A CH-47 Chinook drops off soldiers and equipment into the mountains near Kohe Sofi in Afghanistan, March 28, 2003. (U.S. Army)

returned fire in the breezeway," Neil said. "It was a mental spur—after we heard the words 'let's go,' everything just kind of kicked in."
Moments later, though, the team became separated in the confusion, but with the situation desperate for the Special Forces soldiers, who found themselves facing a determined and larger-than-expected enemy, Pryor and one of his teammates kept moving forward, room to room. They began to enter a room together, but another enemy soldier outside the room distracted the team member, so he stayed outside to return fire.

Pryor first encountered an enemy that was charging out of the room and assisted in eliminating him. Then, without hesitation, Pryor moved ahead into the room and found himself alone with three more enemy soldiers. According to Pryor, the next two enemies he saw were firing their weapons out of the back of the room at his men that were still outside the compound. "I went in, and there were some windows that they were trying to get their guns out of to shoot at our guys that hadn't caught up yet," he said. "So I went from left to right, indexed down and shot those guys up. I realized that I was well into halfway through my magazine, so I started to change magazines. Then I felt something behind me, and thought it was (one of my teammates)—that's when things started going downhill." Pryor said it was an enemy soldier, a larger-than-normal Afghan, who had snuck up on him. "There was a guy back behind me, and he whopped me on the shoulder with something, and crumpled me down." Pryor would later

One of many caves on the list of objectives that the soldiers of A Company, 2nd Battalion, 504th Parachute Infantry Regiment, 82nd Airborne Division, cleared during a combat patrol in the mountains of Adi Ghar, Afghanistan, as part of Operation Mongoose on January 30, 2003. (U.S. Army)

One of several tunnels discovered by Marines of Battalion Landing Team, 3rd Battalion, 6th Marines, 26th Marine Expeditionary Unit (Special Operations Capable), found beneath adobe structures hiding the tunnels. The structures were destroyed and the tunnels collapsed by combat engineers. (USMC)

learn that he had sustained a broken clavicle and a dislocated shoulder during the attack.

"Then he jumped up on my back, broke my night-vision goggles off and starting getting his fingers in my eyeballs. I pulled him over, and when I hit down on the ground, it popped my shoulder back in." Pryor said that after he stood up, he was face to face with his attacker. Pryor eliminated the man during their hand-to-hand struggle. Pryor had now put down all four enemies, but the fight wasn't over yet. "I was trying to feel around in the dark for my night-vision goggles, and that's when the guys I'd already killed decided that they weren't dead yet." Pryor said that it was then a race to see who could get their weapons up first, and the enemy soldiers lost. He then left the room and rejoined the firefight outside. When the battle ended, 21 enemy soldiers had been killed. There were no American casualties and Pryor had been the only soldier injured.

During a search and destroy mission in the Zhawar Kili area on January 14, 2002, U.S. Navy SEALs (SEa, Air, Land) found valuable intelligence information, including this Osama Bin Laden propaganda poster located in an Al Qaeda classroom. In addition to detaining several suspected Al Qaeda and Taliban members, SEALs also found a large cache of munitions in numerous caves and aboveground structures. The SEALs destroyed more than 50 caves and 60 structures by using on-ground explosives and air strikes. (USN)

British newspaper reports regarding the activities of British forces in Afghanistan are rather sparse in detail but nonetheless two actions were notable. The British SAS was involved in two major battles, overcoming enemy fighters who charged at the British from their fortified positions. During the attack by about 100 soldiers on an Al Qaeda training camp/cave complex in Kandahar, officers were involved in close-quarters gun battles and hand-to-hand combat in which knives were used. Enemy forces numbered approximately 200 TAQ fighters. Ultimately, the SAS killed 27 fighters and captured another 30. The TAQ fighters were dug into networks of bunkers, caves and gullies commonly found throughout eastern Afghanistan. According to several newspaper accounts the ferocity with which they fought took the SAS by surprise. The TAQ fighters showed "no regard for their own safety" (Robert Winnett, *The Times*, December 12, 2001). During another action in the Tora Bora complex of the White Mountains scores of Al Qaeda fighters were killed in desperate fighting.

Defensive operations

The Mujahideen/Taliban/Al Qaeda

A declassified study commissioned by the Central Intelligence Agency on the lessons learned in Afghanistan serves as an excellent example of the type of warfare waged both against the Soviet and Coalition forces there by the Mujahideen.

Many of the strengths of the Mujahideen are intrinsic to insurgency warfare. That is, their conduct of the war is very similar to that used historically by insurgency groups in Vietnam or Algeria. The individual Mujahideen's best strength derives from his intimate knowledge of the terrain. Their local militias have tactical mobility, meaning that they could group together as a small unit or disband to fight another day. The element of surprise they gain from their ability to travel and successfully operate at night has given them another advantage. During the jihad against the Soviets, the guerrillas could easily call on other fighters to replenish their numbers. Additionally, the report specifies that the fighters were "unusually rugged and highly motivated." It is further stated that this dedication is derived from a historic dislike of all foreign invaders as well as the fighters' religious fervor. This type of motivation was hurled against the communist government of Afghanistan, and later against the Soviet Army. Furthermore, the Mujahideen enjoyed, like most guerrillas, popular and local support. Pakistan provided sanctuaries, training and financial support—factors that would in today's overly nomenclature-driven military be termed as force-multipliers.

Darunta guerrilla training camps. Excellent satellite imagery from IKONOS that clearly shows cave entrances and training camps. The IKONOS satellite weighs about 1,600lb. It orbits the Earth once every 98 minutes at an altitude of approximately 423 miles (681km). IKONOS can produce one-meter imagery of the geography. The digital imaging sensor is designed to produce images with superior contrast, spectral resolution and accuracy. (Space Imaging)

Along with strengths come weaknesses. The most prominent being a lack of unified leadership or strategy. Short-term strategic planning did exist, i.e. expel the Soviet invader, but the framework did not exist for longer-term strategy. As the Mujahideen were drawn from dozens of different ethnic groups, sporadic inter-fighting did occur, although this was usually limited due to the more pressing presence of an occupying foreign military. Although some fighters did receive occasional training, most did not, and thus no genuine training base could be cultivated. As with all insurgents, the Afghan fighters lacked the massive firepower capability to match that of the government or the occupying army. Thus they relied mostly on small arms as well as shoulder-launched missiles. As the war progressed, however, more forces could draw on small mechanized units. During this time they also developed several cave complexes close to the Pakistani border. Although proximity to Pakistan and Iran was helpful, long and arduous treks had to be made in order to replenish their caches. It was for many years a logistical struggle to maintain adequate war materiel. Given these factors, communication systems were extremely limited, a disadvantage which proved to be extremely helpful as Soviet forces in general were unable to intercept radio transmissions, since few existed.

Their lack of unity or a sophisticated command network turned out to serve the resistance well. Soviets and communist Afghan army units were hard pressed for intelligence and were equally impotent to penetrate the Mujahideen. The study argues that "flamboyant individual actions, dispersal into small elements, decentralized leadership and spontaneous operations decreased tactical predictability and presented few large-scale targets for the Soviets and the government army."

There were some exceptions to this type of ad hoc warfare; for example, the highly experienced Mujahideen who frequently bloodied the noses of elite Soviet troops that penetrated deep into the mountains.

The CIA report does insist that the Mujahideen did not have any kind of universal tactic. Indeed, it argues that most tactics were developed locally through trial and error. What may have worked for a unit defending the heavily developed tunnel systems at Zhawar Kili may not have worked for those defending Tora Bora or any other area. The report essentially explains that good tactics or bad tactics depend on the area, the group, its leader and luck. No standard operating procedures existed. It does not, however, dismiss the improvements that would naturally develop during any war. It concludes: "Mujahideen success was closely linked to the amount of weapons and ammunition they had at their disposal. This varied considerably in different areas and at different times. Nonetheless, the Mujahideen validated several basic tenets of insurgency warfare against overwhelmingly superior conventional forces." The fighters are described as "fiercely independent, basically uneducated peasant or urban fighters in a diverse or political environment."

The report, authored as a tool for American personnel to work with these groups, points out that the greatest mistake the Mujahideen could make would be to form larger groups under unified commands and that American teams ought to continuously be on alert to avoid this from happening.

The manual points out other lessons, including:

- Know the enemy's tactics and routines, as it will allow insurgents to predict enemy tactics. In the case of Al Qaeda, one can argue that they are intimately familiar with U.S. procedures as they were involved in earlier fights with Somalis against U.N. and U.S. forces.
- Be aware of personal and group rivalries and conflicts as they can lead to problems within Mujahideen groups.
- Take full advantage of the cultural and linguistic identities between insurgent and government personnel.

Defending a cave complex

Defensive operations of cave/bunker/tunnel complexes would remain relatively the same throughout 1979–2004, with the notable exception of better technology available regarding shoulder-launched weapons such as the American-manufactured Stinger missile that has had an excellent rate of success in downing Soviet aircraft. One other aspect in setting up defensive perimeters and positions is that one inevitably gets better at it through trial and error and certainly the Mujahideen received plenty of opportunities to improve themselves. After all they faced 80,000–115,000 Soviets in country, as well as another 250,000 communist Afghan Army forces.

During the jihad against the Soviets and most currently the Coalition forces, Mujahideen/TAQ fighters have always been cruelly exposed to relentless aerial attacks. Fixed wing and helicopter attacks have caused great consternation and casualties. Because of this, the militias tend to travel at night and in small groups. As NOGs (night vision goggles) have improved the ability of soldiers to wage war on a 24-hour basis, the fighters have also sought better means to

Images of the Tora Bora region from IKONOS. It is with this type of technology that Coalition forces, led by the U.S., attempted to destroy the hard-core elements of Al Qaeda, including Osama bin Laden, during Operation Enduring Freedom. The combination of high technology with small forces is the brainchild of the Secretary of Defense Donald Rumsfeld and led to numerous failed campaigns in the "Global War on Terrorism." (Space Imaging)

defend themselves and have adapted several other stratagems of defense. Although heavy machine guns and missiles were employed even at the beginning of the conflict, the simplest of countermeasures were put into place: dispersal, cover and terrain knowledge. Afghanistan's natural fortress has been instrumental in the survival of the Mujahideen.

Another lesson learned from the Soviet War in Afghanistan, as identified by the unclassified CIA report, is to "position base camps [such as Zhawar Kili] so as to reduce the effectiveness of air attacks." A typical base camp would be "tucked into a crease between two ridges of a mountain with steep slopes rising on three sides. Heavy machine guns are placed near the crests of surrounding hills in emplacements chiseled out of the rock, with bomb shelters to protect gunners." Many more emplacements were built to allow for the movement of guns. Ultimately, as the war progressed, more weapons became available and to this day it is not unusual to see unmanned or abandoned weapons systems dotting the mountains. A base camp would also have shelters for ammunition and personnel. Most of this construction was carried out with simple jackhammers or other mining tools.

When looking at the IKONOS satellite image taken in 1999 of the Darunta complex, one can make out the various tunnel entrances. The complex featured several training camps, including Abu Khabab's, the Assadalah Abdul Rahman camp, run by the son of the blind cleric Omar Abdul Rahman, who is now jailed in the United States for the explosion at the World Trade Center in 1993. Pakistani fighters seeking to annex Kashmir are located at the Hizbi Islami Camp. Defensive trench lines dot some of the ridges as well as observation posts, and dozens of tunnel entrances are visible, as well as a possible helicopter landing pad and road blocks or check points. Most of these training/base camps also house family members.

In the event of an air strike, the release bombs or rockets would have to be dropped after passing the first mountain and thus would likely land over the site at the bottom of the crease and strike the opposite hill. If an aircraft or helicopter decided to run straight at the crease, it would have to fly though a barrage of interlocking fields of heavy machine-gun fire, coupled with shoulder-launched missiles. Even if the craft survived the bomb run, the very nature of the tunnel construction would require a direct hit to cause any serious damage. With the dawn of thermobaric weapons, standards of precision have become less demanding.

Mujahideen fighters have used a variety of different types of weapons to safeguard their defensive positions against Soviet and Coalition forces. Numerous portable shoulder-fired missiles have made up their growing arsenal. These include the old (usually Egyptian-made) Strela-1, Strela-2, and Strela-2M (upgraded) missiles as well as American Red Eye, Stinger and British Blowpipe missiles. Though the Blowpipes have been rather unsuccessful. Even especially trained Pakistani advisors who helped fight the Soviets and their surrogate Afghan forces at Zhawar Kili were unable to use the British missiles with any degree of accuracy.

Timothy Gusinov, a Soviet veteran who served two tours of duty, a total of four to five years, in Afghanistan notes that, according to information gathered from POWs, the Blowpipe performance was disappointing because of its low accuracy, heavy weight, and complicated guidance system. Blowpipes were used en masse during the 1986 assault on Javara south of Khost. Gusinov states: "I personally witnessed two to three simultaneously launched Blowpipe missiles missing a single aircraft and exploding in the air." (*Soviet Special Forces (Spetsnaz): Experience in Afghanistan*, Timothy Gusinov.)

Other types of antiaircraft weapons include the Soviet-made 12.7mm DShK M1938/46 and NSV, and 14.5mm ZSU-1, 2, and 4 (indicating the number of barrels) heavy machine guns. Other than the NSV, these weapons were also procured from China and Egypt. Besides employment as ground-mounted antiaircraft guns these weapons were used by the Soviets as tank and armored

Bunker/pillbox

Bunker/pillbox

One of the most difficult obstacles for an invading army to overcome in mountainous terrain are bunkers. Although simple in design they tend to be relatively well camouflaged and resistant to small-arms fire. Airpower has been the answer for Soviet troops as well as Coalition units in their campaigns to root out guerrillas from 1979 to 2004. In certain areas of Afghanistan a series of bunkers and caves cut into the side of mountains can prove deadly to aircraft due to the well-planned interconnecting fields of fire. One heavy machine gun that packed a good punch against aircraft was the 12.7mm Soviet DShK 1938/46. It is not uncommon to fly over Afghanistan and spot a number of camouflaged "abandoned" machineguns.

personnel carrier-mounted guns. They had an effective range of 5,000 (1,500m) to 6,000ft (1,800m) and their armor-piercing-incendiary ammunition was effective against aircraft. Even after the introduction of surface-to-air missiles, the 12.7mm and 14.5mm machine guns caused from 50 to 70 percent of helicopter losses and damage, and from 40 to 50 percent of aircraft losses and damage. Also in limited use were Swiss 20mm Oerlikon antiaircraft guns and the Soviet-made self-propelled, four-barrel 23mm ZSU-23-4 automatic gun system known as *Shilka*, which is still used by the Taliban and the Northern Alliance.

For better protection of their fortified bases and strongholds, enemy forces established a local early warning system that consisted of a net of observation posts. Small radio stations were established as far as 3 to 9 miles from each post. This distance does not seem like much when flying in a jet, but it is enough to give advanced warning of approaching helicopters. Also, such posts kept air force bases under observation, reporting every group takeoff. To counter such a net, striking teams should take a deceptive course, and then revert to their planned course once out of the observation area.

Enemy air defense of fortified bases began around 2½–3½ miles out from the main base area. Air defenses included heavy antiaircraft machine guns and occasional surface-to-air missiles located on high mountain ridges. The concentration of air defenses gradually increased toward the center of main bases and fortified areas. The number of heavy machine guns defending a base varied depending on its size and importance, but could range from 60 to 80 pieces in a particular area. Crews were tough. Often, when a gunner was killed or wounded, another trained crewmember immediately replaced him.

Soviet pilots nicknamed the antiaircraft machine guns "welding machines," because from the air the flashes that occurred when they were fired reminded the pilots of welding works in progress. Fortified areas with large numbers of antiaircraft machine guns were called "welding workshops."

Special "free-hunting" missile teams were usually composed of 10 to 20 soldiers, one or two trained missile men, and two to three soldiers to carry additional tubes. Other team members carried infantry weapons for protection and cover. Hunting teams which operated near air bases, and missile teams which defended near enemy bases, included 4 to 10 member groups whose mission was to kill or capture downed pilots. Pilots' messes at airbases, such as at Bagram and Kabul, were specific targets for mortar or rocket barrages. Sometimes such teams would climb to incredible heights in order to attack or engage transportation aircraft that the Soviets thought were flying at safe altitudes.

From 1979 to 1989 the Soviets and allied Afghan forces attempted to seize cave complexes. A ground element would follow an air attack, but to attempt an attack uphill was tremendously disadvantageous. U.S. doctrine during Operation Enduring Freedom called for at least a 3:1 attack ratio, although under the U.S.'s high-tech war, fewer ground troops have been employed. In any event, the Soviets did make adjustments, as their general strategic outlook was to avoid casualties. The preferred method of attack was via a heliborne air

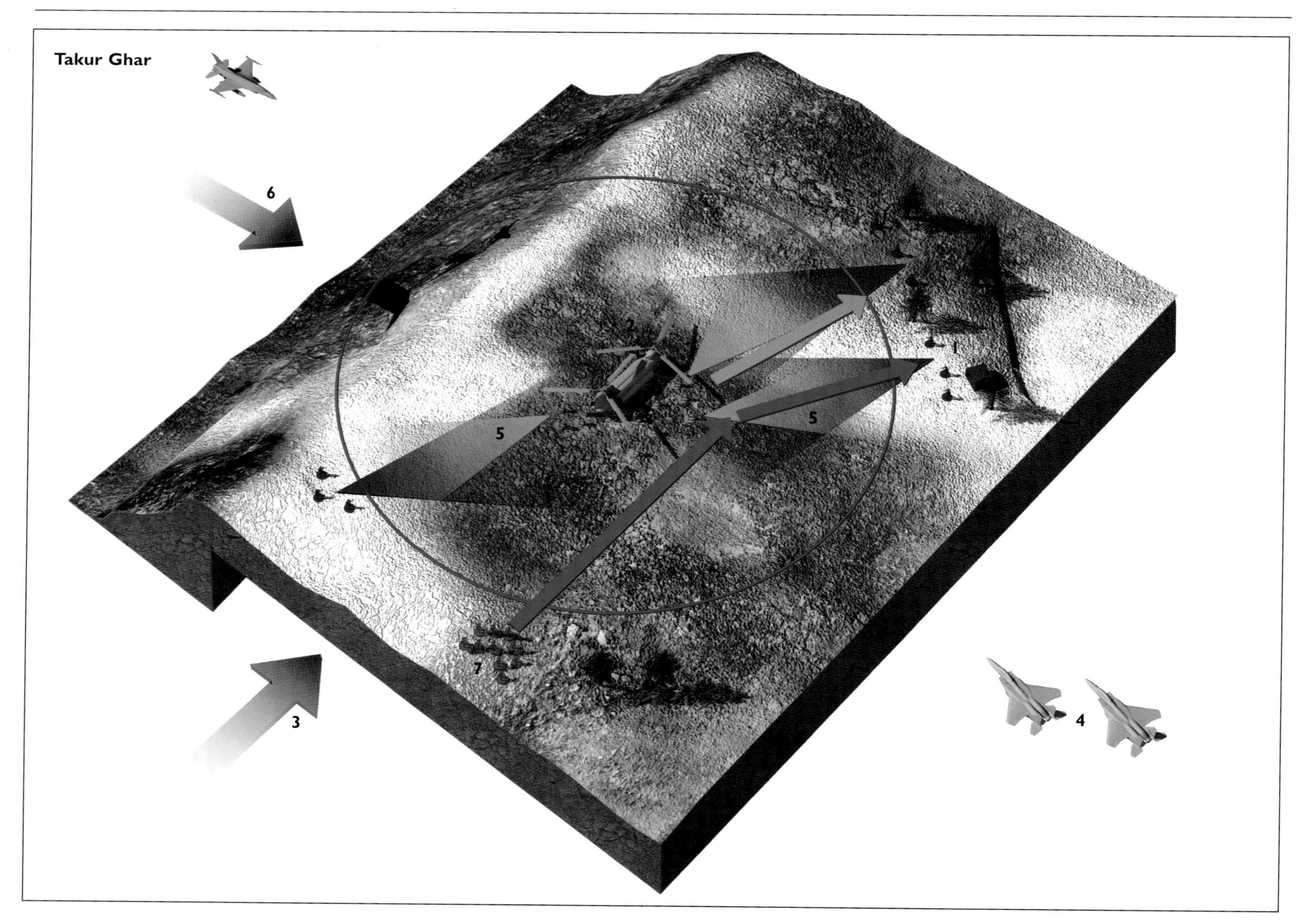
Takur Ghar
6
2
1
5
5
7
3
4

Takur Ghar

1) The mountaintop of Takur Ghar was occupied by a small number of TAQ fighters who had built a bunker (as well as a number of shallow trenches.

2) The guerillas managed to engage several transport helicopters ultimately downing one within 75 yards of their defensive positions.

3) Although only a handful of fighters were present at Takur Ghar, others were spread throughout the area attempting to reinforce the Mujahideen.

4) The American Special Operations Force (SOF) was nearly over-run had it not been for pinpoint aerial strikes by F-15 and F-16 combat aircraft (targets outside of the Blue perimeter). Although the Mujahideen poured massive amounts of small-arms fire, including low-tech RPGs responsible for damaging numerous American helicopters, they were ultimately no match for the sophisticated combined-arms tactics SOF was able to coordinate.

5) Initially, the SOF unit returned fire to three central areas: 12 o'clock, 2 o'clock and 8 o'clock. The Mujahideen forces on the hilltop probably numbered no more than a reinforced squad (12–14).

6) Other fighters engaged the beleaguered team sporadically from other mountains and very small groups of local militia attempted to reinforce the rather thinly stretched TAQ unit on Takur Ghar. TAQ reinforcements were engaged by other Coalition forces, who received their target packages from SOF reconnaissance teams surrounding the immediate area.

7) Another group of Rangers successfully climbed the mountain after commandos from SEAL Team 6 were beaten back. The Rangers successfully assaulted the handful of TAQ fighters, but still had to wait for a nighttime extraction as Coalition forces at the time were engaged in an extremely difficult operation (Anaconda) and resources were overextended.

assault insertion. Soldiers would be dropped off on landing zones near ridges or mountaintops close enough to the intended target, encircling it to the best of their ability and then seep down toward the camp, reversing the action of the older tactic. The Mujahideen in turn would adjust their tactics, and to safeguard against assaults from above against the base camp would place smaller camps throughout an area, thereby creating better interlocking fields of fire and support systems if possible. An assault on one camp would alert the groups in surrounding areas that would then rush to the defense or launch a counterattack. This type of tactic was employed during Operation Anaconda with great success, ultimately resulting in the use of the first thermobaric bomb by Coalition forces. And customary to successful patterns, if the attacking force was deemed too powerful, the Mujahideen would disperse into small groups of three to four men and slip away from the contested area. A report describes as "particularly gutsy" the uncanny ability of fighters to freeze whenever any reconnaissance aircraft were spotted in the vicinity, blending in with the surrounding terrain. With the advent of thermal imaging devices, this particular tactic of avoiding detection may have run its course.

The Soviet objectives of nearly two decades ago are still relevant to the mission assigned to Coalition forces today in Afghanistan:

- Control the cities and towns.
- Secure the major lines of communications (LOC).
- Train and equip government forces.
- Eliminate insurgent (terrorist) centers.
- Separate insurgents from the population.
- Deny by interdiction outside aid and sanctuary.

The Soviets failed in most of these tasks. Today the Coalition forces are struggling to accomplish the same objectives with one notable difference; Pakistan, though struggling with her own insurgency problems, is working with the U.S. to accomplish the last mission task at which the Soviets failed so miserably.

The unclassified CIA report also draws lessons that may well be applicable to today's armed forces:

- Decentralize planning and execution of low-level planning.

The theory of American combined arms – Battlefield Operating Systems (BOS)

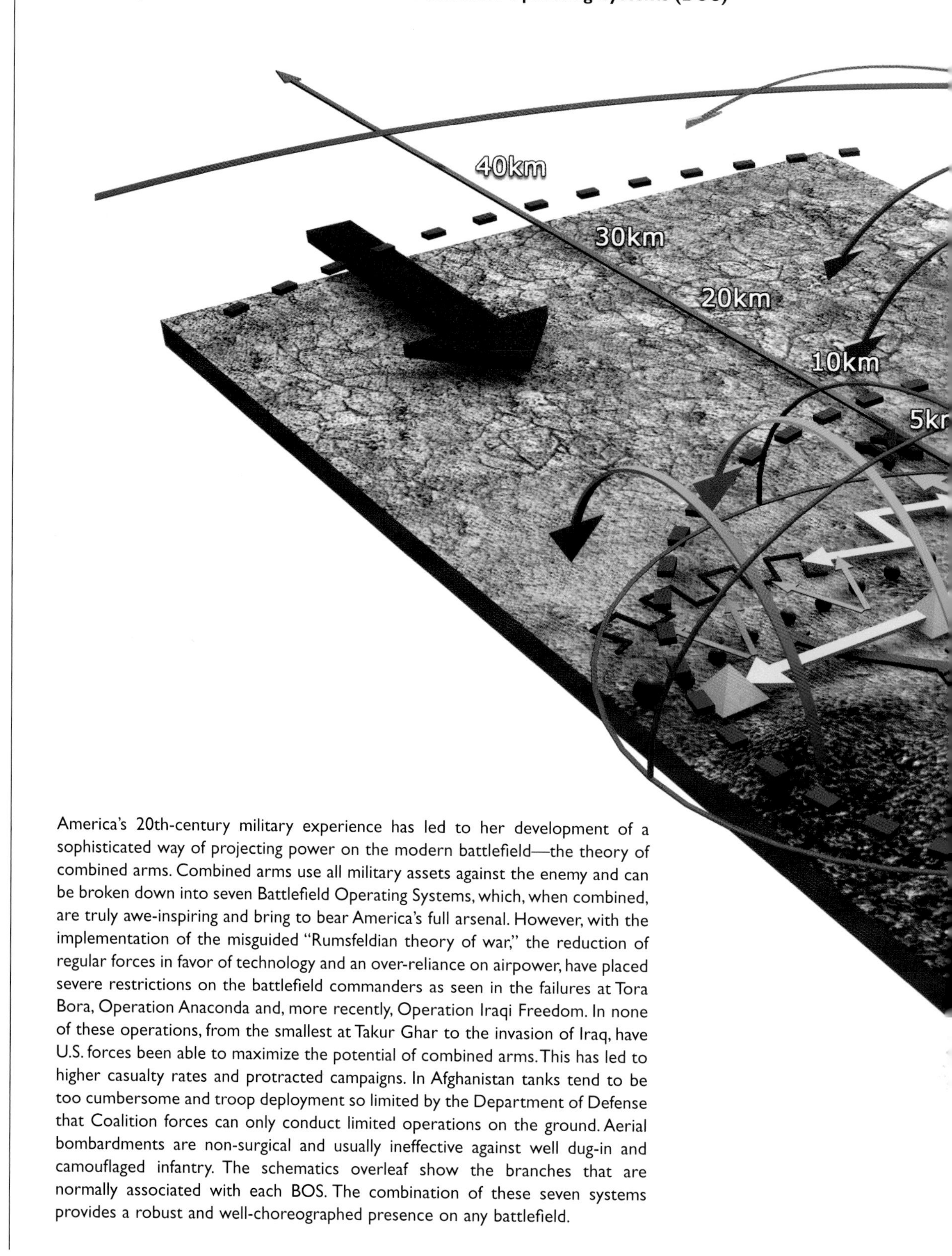

America's 20th-century military experience has led to her development of a sophisticated way of projecting power on the modern battlefield—the theory of combined arms. Combined arms use all military assets against the enemy and can be broken down into seven Battlefield Operating Systems, which, when combined, are truly awe-inspiring and bring to bear America's full arsenal. However, with the implementation of the misguided "Rumsfeldian theory of war," the reduction of regular forces in favor of technology and an over-reliance on airpower, have placed severe restrictions on the battlefield commanders as seen in the failures at Tora Bora, Operation Anaconda and, more recently, Operation Iraqi Freedom. In none of these operations, from the smallest at Takur Ghar to the invasion of Iraq, have U.S. forces been able to maximize the potential of combined arms. This has led to higher casualty rates and protracted campaigns. In Afghanistan tanks tend to be too cumbersome and troop deployment so limited by the Department of Defense that Coalition forces can only conduct limited operations on the ground. Aerial bombardments are non-surgical and usually ineffective against well dug-in and camouflaged infantry. The schematics overleaf show the branches that are normally associated with each BOS. The combination of these seven systems provides a robust and well-choreographed presence on any battlefield.

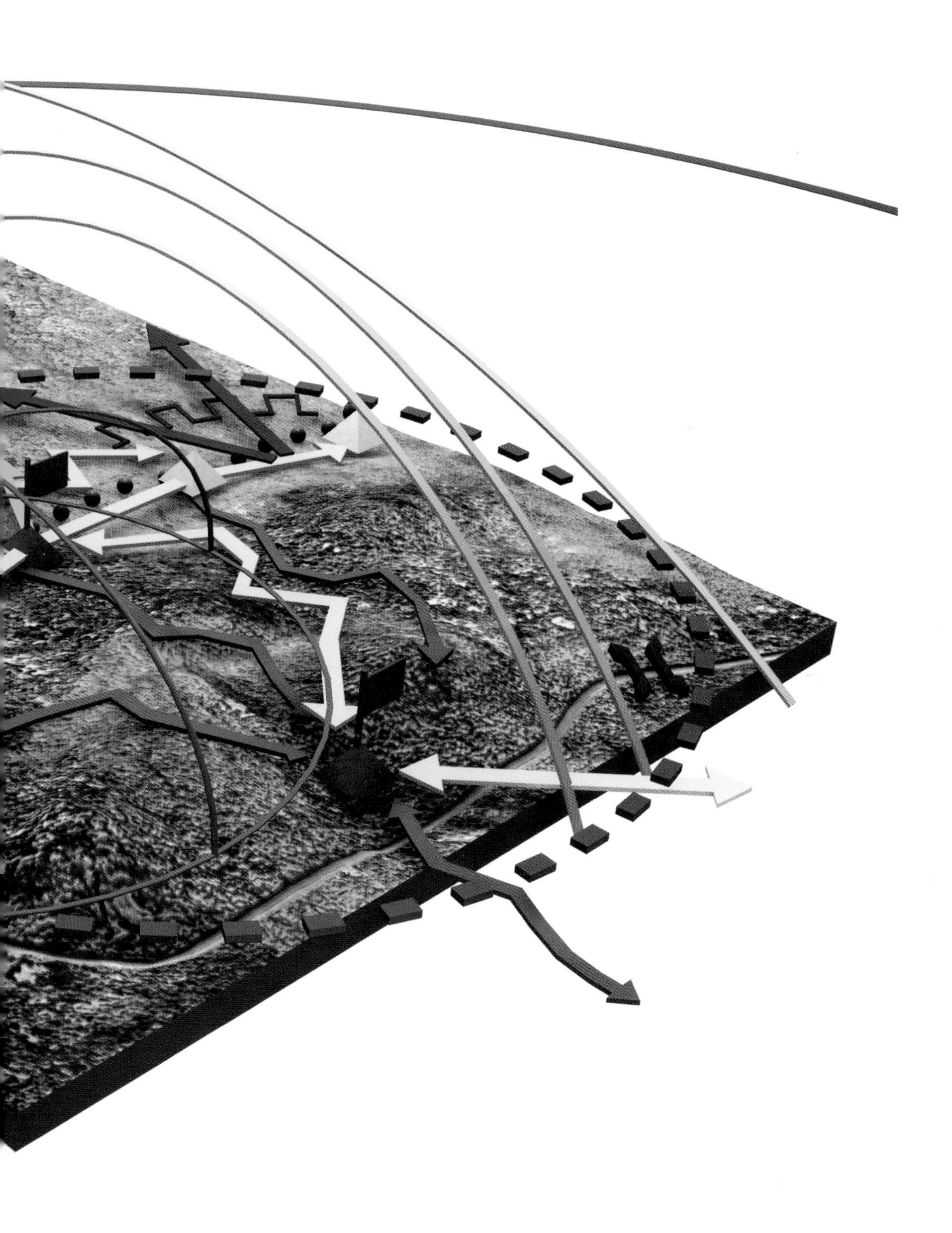

The theory of American combined arms – Battlefield Operating Systems (BOS)

1. **Maneuver** (Infantry/Armor/Special Forces/Aviation)
This BOS constitutes the movement, positioning, and massing of forces on the battlefield.

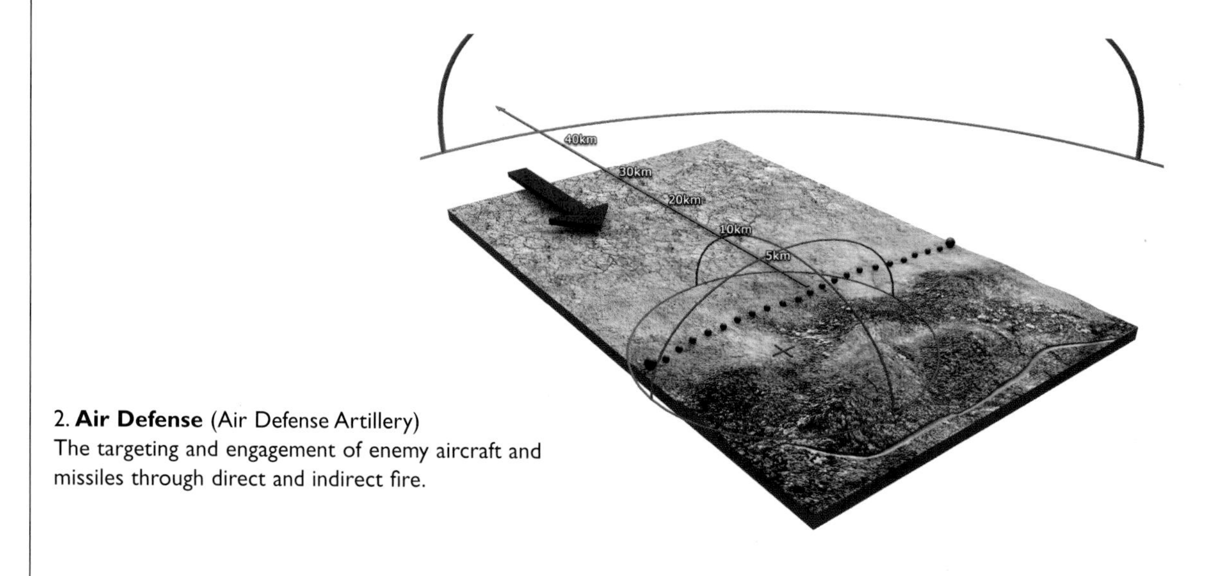

2. **Air Defense** (Air Defense Artillery)
The targeting and engagement of enemy aircraft and missiles through direct and indirect fire.

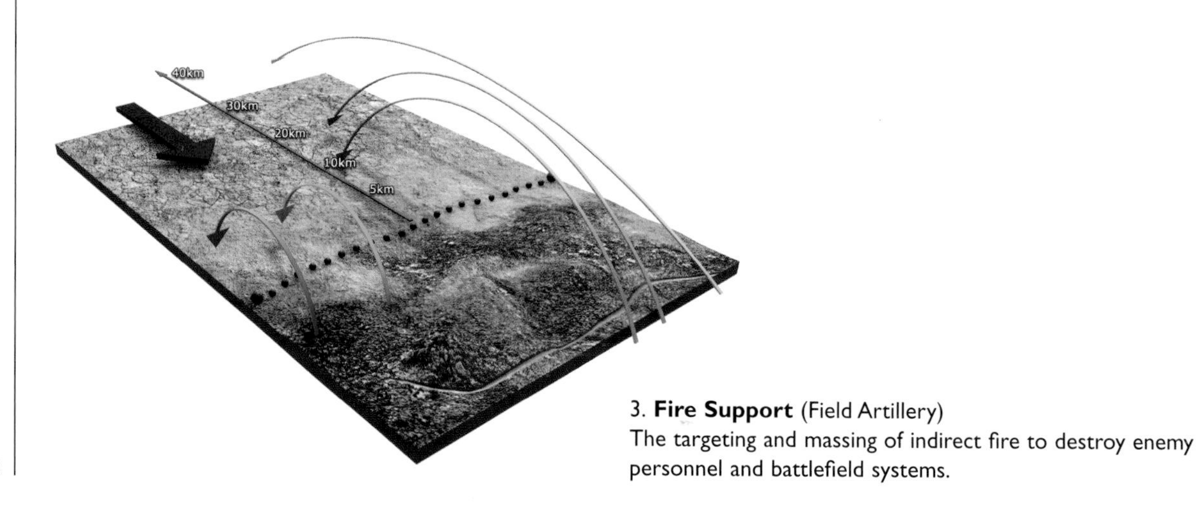

3. **Fire Support** (Field Artillery)
The targeting and massing of indirect fire to destroy enemy personnel and battlefield systems.

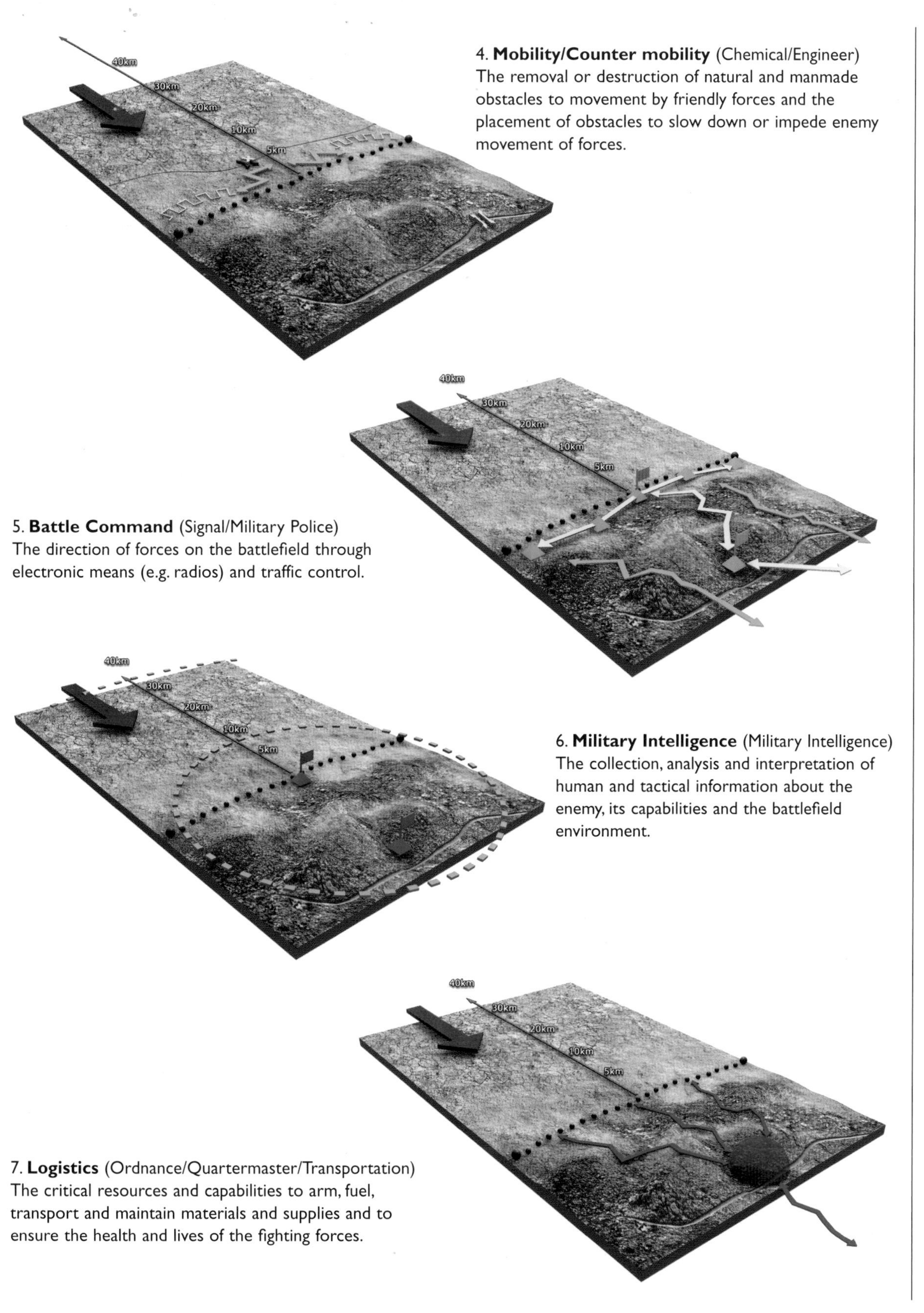

4. **Mobility/Counter mobility** (Chemical/Engineer)
The removal or destruction of natural and manmade obstacles to movement by friendly forces and the placement of obstacles to slow down or impede enemy movement of forces.

5. **Battle Command** (Signal/Military Police)
The direction of forces on the battlefield through electronic means (e.g. radios) and traffic control.

6. **Military Intelligence** (Military Intelligence)
The collection, analysis and interpretation of human and tactical information about the enemy, its capabilities and the battlefield environment.

7. **Logistics** (Ordnance/Quartermaster/Transportation)
The critical resources and capabilities to arm, fuel, transport and maintain materials and supplies and to ensure the health and lives of the fighting forces.

May 6, 2002. A burial mound rests in the middle of the village of Markhanai, which lies in a valley in the Tora Bora region. This region was the last known site where Osama Bin Laden was located. Targets in Tora Bora were bombed with 2000lb GPS-guided Joint Direct Attack Munitions (JDAMs) with hopes of taking out Bin Laden and his followers while closing caves. (USAF)

- Train battalion commanders and staffs to operate with combined arms task forces.
- Use dismounted infantry or air assault teams to identify and destroy antitank ambushes.
- Constantly review and reassess the effectiveness of tactics being applied against insurgent forces.
- Use smoke as a defensive measure when caught in an ambush.
- Helicopter operations at night require extensive training and familiarity with the terrain.
- Use air transport whenever possible.
- Develop countermeasures and train to defend against chemical agents under special conditions (mountain, desert, arctic terrain).
- Prepare troops to encounter flame/incendiary weapons.

The general conclusion drawn from the Soviet experience is that "limited goals mean limited commitment."

Secretary of Defense Donald Rumsfeld looking for answers from a higher authority. (DoD)

The sites at war

The Battle for Zhawar Kili

The story of the Battle of Zhawar Kili, based upon the excellent article entitled *The Campaign for the Caves; the Battles for Zhawar in the Soviet-Afghan War* written by Lester W. Grau and Ali Ahmad Jalali, provides a vivid illustration of the Soviet's experience amidst the unique terrain of Afghanistan.

Zhawar was a Mujahideen logistics transfer base in Paktia Province in the eastern part of Afghanistan. It was located 2½ miles from the Pakistan border and 9 miles from the major Pakistani forward supply base at Miram Shah. Zhawar began as a Mujahideen training center and expanded into a major Mujahideen combat base for supply, training and staging. The base was located inside a canyon surrounded by Sodyaki Ghar and Moghulgi Ghar mountains. The canyon opens to the southeast facing Pakistan. As the base expanded, Mujahideen used bulldozers and explosives to dig at least 11 major tunnels into the southeast-facing ridge of Sodyaki Ghar Mountain.

The first battle

The Mujahideen "Zhawar Regiment," some 500 strong, was permanently based at Zhawar Kili. This regiment was primarily responsible for logistics support of the mobile groups fighting in the area and for supplying the Islamic Party (HIK) groups in other provinces of Afghanistan. The regiment had a Soviet 122mm D30 howitzer, two tanks (captured from the DRA post at Bori in 1983), some twelve-tube Chinese-manufactured 107mm BM-12 multiple rocket launchers (MRL) and some machine guns and small arms. A Mujahideen air defense company defended Zhawar with five ZPU-1 and four ZPU-2 antiaircraft heavy machine guns. These 14.5mm air defense machine guns were positioned on high ground around the base.

The defense of the approaches to the base was the responsibility of Mujahideen groups from the National Islamic Front of Afghanistan (NIFA), the Islamic Revolutionary Movement (IRMA), and the two Islamic Party factions (HIH and HIK). There were six major Mujahideen supply routes into Afghanistan. Twenty percent of all the Mujahideen supplies came through the Zhawar route. The overall Mujahideen commander of Paktia Province, including Zhawar base, was Jalaluddin Haqani, who was a member of HIK.

In September 1985, the DRA attacked, but after 42 days of fighting and low on ammunition and supplies, withdrew. Mujahideen casualties were 106 killed in action and 321 wounded in action. DRA and Soviet losses were heavy, but their numbers unknown since they had evacuated their dead and wounded. Zhawar was a symbol of Mujahideen invincibility in the border area, and the Soviets and DRA felt that they had to destroy this myth. Field fortifications around Zhawar were neglected and incomplete, but the excellent field defenses at the mouth of the Manay Kandow Pass bought time to improve the other fortifications. Their complacent attitude almost cost the Mujahideen their base, and it was only the unexpected appearance of Mujahideen armor at a crucial minute that prevented a DRA victory.One principal advantage that the Mujahideen had was their ablility to move men and supplies from Pakistan throughout the battle.

The second battle

In February 1986, the Soviets felt that the DRA should now take the leading combat role against the Mujahideen and urged the DRA to again attack and destroy Zhawar.

On February 28, government forces, covered by Soviet aviation, began the move out of Gardez to the combat zone. However, when units of the Afghan force arrived in the Matwarkh region, they ceased further movement and stayed there for about a month, simply marking time.

Taking advantage of this passive government force, the Mujahideen began to launch shelling attacks against them. The weather was wet snow mixed with rain and a strong wind. After several days, the composite force moved into the valley and prepared for the offensive.

The initial disaster

Several divisions (under-strength) prepared to advance from the east and the west. Sometime around midnight on April 2, the DRA began a two-hour artillery and aviation preparation of the target area. Then six armed helicopter transport ships flew from Khost airfield to insert the initial assault group of a Commando Brigade. The commandos landed without opposition, but the ground assault ran into immediate, heavy resistance from Mujahideen defending the Dawri Gar Mountain. The ground advance was forced to halt.

The commandos had landed some 3 miles (5km) inside Pakistan, beyond the base at Zhawar. As the command post informed them of this circumstance, the insertion group commander quietly answered, "I understand. We will withdraw." But after an hour he reported that he was surrounded and locked in combat.

The air assault was botched and consequently made the situation much worse. The rest of the brigade was committed to combat—not onto the Dawri Gar Mountain landing zone, which was well populated with Mujahideen, but onto the open areas around Zhawar itself.

One Mujahideen commander saw approximately 20 transport helicopters flying over and radioed the commanders at Zhawar to warn them. After his radio message, he saw another group of helicopters, including some heavy transport helicopters, flying in the same direction. These were escorted by jet fighters. He again radioed this information to Zhawar. Zhawar had 700–800 Mujahideen combatants, plus air defense forces, at the time.

The usual Soviet/DRA pattern for an attack on a Mujahideen base was to pound the area heavily with air strikes and then follow the air strikes with air assault landings, artillery fire and a ground advance to link up with the air assault forces. The air strike gave the Mujahideen commanders warning, reaction time and a solid indicator where the attack would go. In this case, the Mujahideen were caught by surprise. Their intelligence agents within the DRA failed to tip them off and the helicopters landed the rest of the Commando Brigade on seven dispersed landing zones around Zhawar. There were 15 helicopters in the first lift, which landed at 0700hrs. More lifts followed, putting the entire brigade on the ground. The first two helicopters landed on Spin Khwara plain. Most of the helicopters landed on the high ground to the west of Zhawar. Mujahideen gunners destroyed many helicopters while they were on the ground. Following the air assault, Soviet jet aircraft bombed and strafed Mujahideen positions. Mujahideen air defense proved to be not very effective against these aircraft.

Instead of defending from positions pounded by fighter-bombers and close air support aircraft, the Mujahideen went on the offensive and attacked the landing zones. They quickly overran four of them and captured many of the DRA commandos. Mujahideen reinforcements moved from Miram Shah in Pakistan to Zhawar and took the commandos from the rear. The commandos were trapped between two forces and were killed or captured. By the end of the day, the Mujahideen captured 530 commandos.

Meanwhile, Soviet aircraft with smart munitions made ordnance runs on the caves. Since the caves faced southeast toward Pakistan, the Soviet aircraft overflew Pakistan in order to turn and fly at the southern face with the smart weapons. Smart missiles hit the first western cave and killed 18 Mujahideen outright. Smart missiles hit the second western cave and collapsed the cave opening trapping some 150 Mujahideen inside. This second cave was 500ft (150m) long and used as the radio transmission bunker. The commander, Jalaluddin Haqani, who had just arrived from Miram Shah, was among those trapped in the second cave.

Other Soviet bombers followed with conventional ordnance. They dropped tons of bombs and, in so doing, blasted away the rubble blocking the cave entrances. The trapped Mujahideen escaped. The battle for the remaining landing zones continued. There was one group of commandos on high ground who held out for three days before they were finally overrun. The chief of counter-reconnaissance in one of the commando battalions managed to exfiltrate and lead 24 of the commandos to the safety of their own forces, a trip that took eight days. Of the 32 helicopters assigned to the mission, only eight survived.

By April 9, the divisions that attempted to link up with the already destroyed commando unit had pulled back to their start points. The DRA continued to fight for the possession of the Manay Kandow Pass for some ten days following the air landing. The Mujahideen attacked the DRA lines of communication and the airfield at Khost, while the Mujahideen holding the Manay Kandow Pass checked their advance.

A plan was formed to reinforce the effort with three DRA regiments, a DRA spetsnaz battalion and six Soviet battalions. Twelve days were allocated to prepare for resumption of the operation.

Once again

Following urgent requests from the leadership of the DRA, five battalions of Soviet forces were sent to Khost and Tani between April 5 and 9. Soviet Forward Air controllers were assigned to work with Afghan Forward Air Controllers in the infantry divisions and the reinforcing Soviet unit commanders were assigned to work with the Afghan division commanders. They reworked the operations plan while the force was refitted. The total DRA/Soviet force now exceeded 6,600 men.

During the refitting, restructuring and replanning efforts, the communists kept the pressure on the Mujahideen with air strikes and artillery. In the first battle for Zhawar, DRA/Soviet artillery and air strikes stopped at night, but this time they were conducted around the clock. At night, they dropped aerial flares for illumination. This heavy fire support continued for 12 days. The tempo of the air and artillery increased on the morning of April 17.

Pakistan was clearly concerned with the major battle raging on her border. The Mujahideen lacked effective air defense against helicopter gunships and the strafing and bombing attacks of high-performance aircraft. The Mujahideen had some British Blowpipe shoulder-fired air defense missiles, but they were not effective. Pakistan sent some officers into Zhawar during the fighting to

British Royal Engineers of Task Force Jacana destroy a cave complex on the border between the Paktika and Paktia provinces in Afghanistan on May 10, 2002, during Operation Snipe. This was reportedly the largest explosion set off by the Royal Engineers since World War II. (DoD)

Zhawar Kili

On January 3, 2002, Coalition aircraft bombed Zhawar Kili, a well-known training center, supply depot and public relations site. U.S. Navy SEALs and Marines conducted Sensitive Site Exploitation (SSE) and in the process destroyed 50 caves and 60 aboveground structures. No enemy personnel were encountered. The area had been bombed in 1998 in response to the U.S. Embassy bombings in Kenya and Tanzania. Zhawar Kili faces southeast toward Pakistan and is roughly 3 miles from the Pakistani border. The camp is composed of an aboveground site and two separate cave complexes, covering an area approximately three square miles. Zhawar Kili has at least 11 major tunnels, with some of them 1,600ft (500m) long, containing a hotel, a mosque, arms depots and repair shops, a garage, a medical point, a radio center and a kitchen. A gasoline generator provided power to the tunnels and the hotel's video player. This impressive base became a mandatory stop for visiting journalists, dignitaries and other "war tourists" during the Mujahideen's war with the Soviet Union. Apparently, this construction effort also interfered with construction of fighting positions and field fortifications.

1) A part of a village usually populated with families of fighters and other noncombatant personnel. Some of the buildings can have tunnel entrances within their structures. A reinforced and well-built entrance leads into the side of a mountain.

2) Tunnel system leading to a number of separate passages containing caves/rooms, some continue to other tunnels. Usually several escape routes exist.

3) Vertical ventilation shafts throughout the complex ensure adequate air supply.

4) Rooms within the structure tend to be primarily arms and ammunition caches but at Zhawar Kili more elaborate uses were made of the complex.

3
3
3
4

take out the attacking aircraft with the British Blowpipe shoulder-fired missiles and train the Mujahideen. After climbing a mountain and firing 13 Blowpipe missiles to no avail, a Pakistani captain and his NCO were severely wounded by the attacking aircraft.

The renewed ground attack began on the morning of April 17. In order to deceive the Mujahideen and divert their forces, the eastern group began its attack at 0630hrs and the western group began at 1030hrs. Multiple attacks on the mountain failed. When the artillery fire preparation started, the Mujahideen sheltered in caves and when the preparation ceased, they reoccupied their firing positions and repulsed the attack. The DRA/Soviet adviser observed this and during the night, they silently moved one of their regiments toward the summit and, at dawn, launched an attack on the Mujahideen without any artillery preparation. The Mujahideen did not expect this and faltered. The regiment captured the summit in a matter of minutes. The Mujahideen fell back from the area into the higher mountains and slowly the DRA/Soviet force moved through the pass.

At the same time, another DRA/Soviet force launched a flanking column that moved to the east. This column moved toward the east flank of Zhawar where a regiment of Mujahideen waited in defense. However, as the DRA column neared, the regiment withdrew without a fight as rumors spread among the Mujahideen that Jalaluddin Haqani was dead. The Mujahideen evacuated Zhawar and moved high into surrounding mountains as the two ground columns closed onto Zhawar.

The Mujahideen were unable to evacuate most of the stores from Zhawar. They pulled out the two T-55 tanks and fought the advancing column for a while before abandoning the tanks in the foothills. The Soviet and DRA forces entered Zhawar at noon on April 19, 1986.

A short stay

A narrow stretch of mountain road led to a 500ft-wide (150m) canyon whose sides stretched upwards for a mile. Caves were carved into the rock face of the side facing Pakistan. The caves were up to 30ft (10m) long, 12ft (4m) wide and 10ft (3m) tall, with walls faced with brick. The cave entrances were covered with brightly painted powerful iron doors. There were 41 caves in all, all with electricity. Behind a fence stood a mosque with a beautiful brick entrance and a hospital with new medical equipment that was manufactured in the United States. There was even an ultrasound machine, which was moved to Khost hospital. There was nickel-plated furniture, including adjustable beds, and a library with English-language and Farsi-language books. There was a bakery and by the entrance was a stack of fresh naan bread. In the storage area, there were metal shelving units where boxes of arms and ammunition were neatly stacked. Further on, there was a storage cave for mines. There was every kind of mine imaginable: antitank, antivehicular and antipersonnel mines from Italy, France, the Netherlands and Germany. Hand grenade and artillery simulators were stored separately. The demolition explosives of various types and detonators were stored in a separate cave. In the very furthest part of the base were repair and maintenance bays complete with grease pits. There was a T-34 tank in one of them. The tank was serviced, fueled and had new batteries. It started right up and drove out of the service bay. Above the storage caves was a beautiful building marked "Hotel." There was overstuffed furniture inside and the floors were covered with carpets. A soldier remarked how many aircraft had worked this site over and the hotel and caves were still intact.

The Afghan soldiers began to loot the base. Even the 2m-high brick facing wall was pulled down and hauled back to Khost. The DRA had no intention of staying in Zhawar long enough for the Mujahideen to organize a counterattack, and the Mujahideen were already moving up to the Pakistan border to fire on the communist forces.

Colonel Kutsenko was in charge of destroying Zhawar and had four hours to do so. He split up the detonation of the caves and buildings between the sappers of the 45th Engineer Regiment of the 40th Army and Afghan sappers. He knew that he could not destroy the caves in the available time. Above the caves was a 100ft-thick (30m) layer of rock. If they could drill a 3–6ft shaft into the ceiling, they could have crammed that full of explosives and caused a collapse, but there was not time to do that before the troops had to leave. So the sappers stacked about 200 antitank mines in the primary caves and rigged them for simultaneous electric detonation. Even if they had placed in ten times more explosives, it would not have made any difference since the force of the explosion would follow the path of least resistance and the caves would channel the force out the caves' mouths.

And the moment finally arrived.

> The caves ... shot out their contents. After the dust settled, the canyon was filled with clumps of earth, shattered bricks and stones. The caves were swept clean, but were somewhat larger and their entries were partially clogged by rockslides from above. The gates were torn pieces of iron laying at the foot of the opposite canyon wall.

The combat soldiers were withdrawing as the sappers remained behind to mine the base. The work was hard and complicated by the lack of time. The sappers had to depart before nightfall. At 1700hrs, the command was given for the remaining force to leave and head for Tani. The Mujahideen were quick to fall on the heels of a retiring foe. Anyone who fell behind or stopped would be in serious trouble. Rockets fired from across the Pakistan border were landing near the sappers and these rounds were becoming more precise. It was time for the sappers to join the exodus. Kutsenko gave the command to depart on his radio. The Afghan sappers immediately quit working and boarded their armored vehicles. The Soviet commander of the sappers answered "Right away." His "Right away" lasted 15 minutes. Kutsenko again called him and ordered that they cease work and depart. Their commander again answered "Right away." Kutsenko then I told him "You may stay here for an hour, but your soldiers need to quickly join the convoy. The Mujahideen are here and we

Lieutenant Christopher Blaha of the 1st Battalion, 87th Infantry Regiment, 10th Mountain Division, gets ready to throw a grenade into a cave, April 6, 2004. (U.S. Army)

A soldier with 2nd Battalion, 27th Infantry, 25th Infantry Division (Light), pulls security for a squad of soldiers climbing the mountain in search of caves, April 4, 2004. (US Army)

OPPOSITE An aerial view of a rugged mountain pass near Gardez. The mountainous terrain provides many hiding places for groups like Al Qaeda and the Taliban. (U.S. Army)

are leaving." This time, the Soviet sappers quit work and immediately boarded their vehicles. Kutsenko ensured that everyone was on board and the trail party left, with Kutsenko returning in the captured tank.

After 57 days of campaigning, the DRA held Zhawar for only five hours. In addition to the standard mines and booby traps, the communist forces planted seismic-detonated mines and sprinkled aerial-delivered butterfly bombs over the area. The Mujahideen returned to Zhawar on the following day. The first to enter the area were killed by seismic mines. The Mujahideen withdrew and fired mortars, BM12 and machine guns into the area to set off the rest of the mines. Then they began the slow process of finding the rest of the mines manually. Since the DRA was only in Zhawar for five hours, they did not manage to destroy the caves, but only collapsed some entrances. Weapons that were stored in some of the caves were still intact and useable.

Mujahideen casualties were 281 killed and 363 wounded. DRA and Soviet losses are unknown, but the Mujahideen reportedly destroyed 24 helicopters, shot down two jets and captured 530 personnel. The Mujahideen held a field tribunal and tried and executed the colonel of the Commando Brigade and another colonel who landed with the brigade to adjust artillery fire. There were 78 other officers among the prisoners. They were given a chance to confess to their crimes from different battles and then all were executed. All the soldiers were given amnesty since they were conscripts who were forced to fight. The amnestied soldiers were asked to perform two years of labor service in exchange for the amnesty. They did their service in logistics, were "reeducated" and released after two years.

Aftermath

The DRA and Soviet reluctance to hold Zhawar for any length of time in order to do a thorough job of destroying the base is a strong testament to the ability of the Mujahideen to threaten their lines of communication. The commanders had no desire to risk being trapped in Zhawar and having to mount yet another operation to fight their way out. Their reserves were committed and the danger was real.

The DRA celebrated the fall of Zhawar as a major victory with parades and medals. But Zhawar was back in full operation within weeks of the attack. Having been trapped in some of the caves, the Mujahideen learned to make connecting tunnels , and shortly had them reopened, in some cases improved and lengthened to 1,500–1,600ft long.

The reconnaissance in force at Zhawar Kili during Operation Enduring Freedom

On January 3, 2002, Coalition aircraft comprising four B-1B bombers, four F/A-18 Hornets and an AC-130 gunship, struck Zhawar Kili. This same complex had previously been struck by U.S. cruise missiles in 1998 in response to the U.S. embassy bombings in Nairobi, Kenya, and Dar es Salaam, Tanzania. Following the bombardment, members of SEAL Team 3 along with Explosive Ordnance Disposal (EOD) personnel conducted a Bomb Damage Assessment (BDA) and Sensitive Site Exploitation (SSE) mission. As the mission continued, a reinforced platoon of Marines was brought in as a security force. The SEALs also requested and received vehicles in order to better cover the enormous area of the cave complex, which covered three by three miles. No enemy personnel were encountered, although a few bloodstains and body parts were found. Airstrikes and other explosive devices destroyed at least 60 aboveground structures and about 80 caves.

U.S. Combined Arms during Operation Anaconda

Operation Anaconda (Operation Veritas for the British, Operation Harpoon for the Canadians) began on March 2, 2002, in the mountainous Shah-I-Kot (Place of the King) region south of the city of Gardez, Paktia Province, eastern Afghanistan. Special Operating Forces from several nations set up strategic reconnaissance observation posts throughout the area. The 10th Mountain Division (Light Infantry) and the 101st Airborne Division (Air Assault)[1], in conjunction with Afghan forces, inserted units into the area of operation, covering more than 60 square miles. Elevation ranged from 8,000 to 12,000ft and the temperature range was 15–20°F, less than ideal conditions for the Coalition soldiers.

The intended target was a group of 500–1,000 TAQ militias, including families, in the Shah-I-Kot Valley. The fighters were entrenched along ridges and caves and nearly overwhelmed the Coalition forces as they unloaded from their transport helicopters. The Afghan forces were repulsed during the early moves of the operation. On March 17, 2002, Operation Anaconda concluded; a total of eight American servicemen had been killed and 82 wounded in action.

The U.S. claimed an unqualified success, but the battle was only not lost due to the overwhelming aerial bombardment, including the first use of a thermobaric bomb, and the continuous close air support provided to the soldiers on the ground. Half of the Apache helicopters used in the operation were damaged by small-arms fire. The 10th and 101st were pinned down, and one battalion from the 10th Mountain Division was extracted the next day. Most of the TAQ militia withdrew to Pakistan, but supposedly hundreds of them had been killed during the operation and their stockpiles of ammunition and arms destroyed.

The Canadian 3rd Battalion Princess Patricia's Light Infantry spent five days clearing enemy positions and searching more than 30 caves and found large caches of ammunition and equipment, collected intelligence documents and maps, and searched a few Al Qaeda who had been killed in the airstrikes. The Canadian infantrymen were extracted by helicopter on March 17 and 18, bringing Operation Anaconda to a close.

1 The 10th Mountain Division is a standard light infantry division and has not been mountain-trained. The 101st is a helicopter air assault division and is not parachute-capable. Both divisions' "Mountain" and "Airborne" designations are traditional.

Afghanistan today

Afghanistan has endured decades of war and has seen a whole generation of children grow up with mines and death. Three years into the invasion of Afghanistan by Coalition forces, few things have changed, although hundreds of caves have been destroyed. Mujahideen are still developing cave systems. Ethnic and political strife are as intense as ever. The president of the country uses U.S. bodyguards and the opium crop has dramatically risen since the ousting of the Taliban. The Soviets were unable to defeat the Mujahideen who moved in and out of Pakistan. The U.S. undoubtedly is hoping that an aggressive Pakistan can close off this kind of access, allowing Coalition forces to destroy the remaining TAQ militias.

Curious local Afghani children sit along a poppy field and watch as coalition forces headed by the U.S. Army Criminal Investigation Division (CID) unearth graves in their village of Markhanai on May 5, 2003, in Tora Bora for Operation Torii. (USAF)

Bibliography and further reading

Books and articles

Bahmanyar, Mir, Warrior 065: *U.S. Army Ranger 1983–2003*, Oxford: Osprey Publishing, 2002

Bahmanyar, Mir, *Naval Commandos: Seals 1983–2004*, Oxford: Osprey Publishing, 2004

Bergen, Peter L., *Holy War, Inc.*, New York: Touchstone, 2001

Borovik, Artyom, *The Hidden War*, New York: The Atlantic Monthly Press, 1990

Cincinnatus, *Operation Anaconda*, Boulder, Colorado: Soldier of Fortune Magazine, July, 2002

Coll, Steve, *Ghost Wars*, New York: Penguin Press, 2004

Dawood, N.J., *The Koran*, New York: Penguin Books, 1974

Evans, Martin, *Afghanistan: A Short History of the People and Politics*, New York: Harper Collins Publishers, 2002

Isby, David, Men-at-Arms 178: *Russia's War in Afghanistan*, London: Osprey Publishing, 1986

Jalali, Ali Ahmad, and Grau, Lester W., *Afghan Guerilla Warfare: In the Words of the Mujahideen Fighters*, St. Paul, MN: Motor Books International, 2001

Kaplan, Robert D., *Soldiers of God: With Islamic Warriors in Afghanistan and Pakistan*, New York: Vintage Books, 2001

Kushner, Harvey W., *Encyclopedia of Terrorism*, Thousand Oaks: Sage Publications, 2003

Nyrop, Richard, and Seekins, Donald M., *Afghanistan Country Study*, The American University, 1986

Rashid, Ahmed, *Taliban: Militant Islam, Oil & Fundamentalism in Central Asia*, New Haven: Yale University Press, 2001

Rashid, Ahmed, *Jihad: The Rise of Militant Islam in Central Asia*, New York: Penguin Books, 2002

Santhanam, Manish, K., and Sreedhar, Sudhir Saxena, *Jihadis in Jammu and Kashmir*, New Delhi: Sage Publications, 2003

Smucker, Philip, *Al Qaeda's Great Escape*, Washington DC: Brassey's, 2004

Waller, John H., *Beyond the Khyber Pass: The Road to British Disaster in the First Afghan War*, Austin, TX: University of Texas Press, 1980

Woodward, Robert, *Bush at War*, New York: Simon & Schuster, 2002

Yousaf, Mohammad and Adkin, Mark, *Afghanistan the Bear Trap: The Defeat of a Superpower*, Havertown, PA: Casemate, 2001

Zaloga, Steven J., and Loop, James, Elite 5: *Soviet Bloc Elite Forces*, London: Osprey Publishing, 1985

CIA, *Lessons from the War in Afghanistan*, CIA De-Classified Report.

Websites

http://www.airpower.maxwell.af.mil/airchronicles/aureview/1985/jan-feb/nelson.html – Air & Space Power Chronicles, Soviet Air Power: Tactics and Weapons Used in Afghanistan, Lieutenant Colonel Denny R. Nelson.

http://www.airpower.maxwell.af.mil/airchronicles/aureview/1986/mar-apr/collins.html – Air & Space Power Chronicles, The War in Afghanistan, George W. Collins.

http://www.cbc.ca/news/indepth/targetterrorism/backgrounders/afghanistan_caves.html – CBC News, The Caves of Afghanistan, Martin O'Malley & Sabrina Saccoccio.

http://d-n-i.net/fcs/wilson_wilcox_military_responses.htm – Defense and the National Interest, Military Response to Fourth Generation Warfare in Afghanistan, Greg Wilcox and Gary I. Wilson.

http://fmso.leavenworth.army.mil/fmsopubs/issues/fuelair/fuelair.htm – Foreign Military Studies Office, A 'Crushing' Victory: Fuel-Air Explosives and Grozny 2000, Lester W. Grau and Timothy Smith, Fort Leavenworth, KS.

http://fmso.leavenworth.army.mil/fmsopubs/issues/arty/arty.htm – Foreign Military Studies Office, Artillery and Counterinsurgency: The Soviet Experience in Afghanistan (CALL Publication #98-17), Lester W. Grau, Fort Leavenworth, KS.

http://fmso.leavenworth.army.mil/FMSOPUBS/ISSUES/groundcombat/groundcombat.htm – Foreign Military Studies Office, Ground Combat at High Altitude, Lester W. Grau, and Lieutenant Colonel Hernán Vázquez, Fort Leavenworth, KS.

http://fmso.leavenworth.army.mil/fmsopubs/ISSUES/minewar/minewar.htm – Foreign Military Studies Office, Mine Warfare and Counterinsurgency: The Russian View, Lester W. Grau, Fort Leavenworth, KS.

http://fmso.leavenworth.army.mil/fmsopubs/ISSUES/zhawar/zhawar.htm – Foreign Military Studies Office, The Campaign for the Caves: The Battles for Zhawar in The Soviet-Afghan War, Lester W. Grau and Ali Ahmad Jalali, Fort Leavenworth, KS.

http://g2mil.com/101st.htm – G2mil, The 101st in Afghanistan.

http://www.armscontrol.ru/atmtc/Arms_systems/Soviet_Arms_main.htm – Moscow Institute of Physics and Technology Center for Arms Control, Energy and Environmental Studies, Marat Kenzhetaev.

http://www.comw.org/pda/0201strangevic.html – Project on Defense Alternatives, Strange Victory: A Critical Appraisal of Operation Enduring Freedom and the Afghanistan War, Carl Conetta.

http://www.prospect.org/webfeatures/2002/03/vest-j-03-21.html – The American Prospect, Mountain Division: Why the U.S. Can't Match the British at High Altitudes, Jason Vest.

http://call.army.mil/ – The Center for Army Lessons Learned (CALL).

http://www.newyorker.com/fact/content/?040412fa_fact – The New Yorker, The Other War, Seymour M. Hersh.

http://www.leavenworth.army.mil/milrev/English/MarApr02/alamanac.htm – U.S. Army Combined Arms Center Military Review, Soviet Special Forces (Spetsnaz): Experience in Afghanistan, Timothy Gusinov.

http://www.geocities.com/usarmyafghangearproblems/ – Official U.S. Army Natick Report on Field Equipment Lessons Learned in Afghanistan.

http://www.geocities.com/glossograph/afghan011125daruntact.html, The Al Qaeda Complex at Darunta, Richard S. Ehrlich.

http://www.suasponte.com – Sua Sponte.

http://www.afa.org/magazine/sept2002/0902anaconda.asp – Air Force Magazine Online, The Airpower of Anaconda, Rebecca Grant.

http://globalsecurity.org/military/world/afghanistan/shahi-khot.htm – GlobalSecurity.org, Afghanistan – Shahi Khot.

http://www.courier-journal.com/localnews/2002/07/04/ke070402s236271.htm – Courier-Journal of Louisville, Kentucky, Battle on Takur Ghar, C. Ray Hall.

http://www.defenselink.mil/news/May2002/d20020524takurghar.pdf – Department of Defense, Executive Summary of the Battle of Takur Ghar.

http://www.csmonitor.com/2002/0801/p01s03-wosc.html – The Christian Science Monitor, Anaconda: A War Story, Ann Scott Tyson.

http://www.leavenworth.army.mil/milrev/English/MayJun02/almanac%20geibel.htm – Combined Arms Center Military Review, Operation Anaconda, Shah-i-Khot Valley, Afghanistan, 2 -10 March 2002, Adam Geibel.

http://www.sftt.org/article07102002a.html – SAS Reveal Narrow Escape from Death, Soldiers for the Truth (SFTT), Ian McPhedran.

http://www.defenselink.mil/news/Mar2002/t03272002_t0327asd.html – Department of Defense, 10th Mountain Division Soldiers Conduct Phone Interviews from Afghanistan.

http://www.defenselink.mil/news/Mar2002/n03272002_200203273.html – Department of Defense, U.S. Troops Describe Clearing Afghan Caves, Bunkers, Linda D. Kozaryn.

http://www.geocities.com/equipmentshop/realmountaindivision.htm – Lessons Learned: Put the "Mountain" Back in the 10th Mountain Division.

http://lcweb2.loc.gov/frd/cs/afghanistan/afghanistan.html – Library of Congress, Afghanistan: A Country Study, Craig Baxter.

http://en.wikipedia.org/wiki/Timeline_of_Afghan_history – Afghanistan Timeline.

http://www.agiweb.org/geotimes/feb02/feature_afghan.html – GeoTimes, Afghanistan: Geology in a Troubled Land, J. Stephen Schindler.

http://www.telegraph.co.uk/news/main.jhtml?xml=/news/2001/12/01/wmil01.xml – Telegraph, Wounded SAS men were in cave raid, Michael Smith.

http://www.agiweb.org/geotimes/feb02/feature_military.html – GeoTimes, Military Geology in a Changing World, William Leith.

http://www.globalsecurity.org/org/news/2001/010928-attack01.htm – New York Post, A Closer Look at Camp Osama, Niles Latham.

Index

Figures in **bold** refer to illustrations